How the Earth Works

Earth Science and the National Curriculum

Series Editor Robin Gill

How the Earth Works

BY

P. Brannland
Boundstone Community College, UK

A. Rhodes

Many thanks for your assistance,

Alan Rhodes

1995
Published by
The Geological Society
London

THE GEOLOGICAL SOCIETY

The Society was founded in 1807 as The Geological Society of London and is the oldest geological society in the world. It received its Royal Charter in 1825 for the purpose of 'investigating the mineral structure of the Earth'. The Society is Britain's national society for geology with a membership of around 8000. It has countrywide coverage and approximately 1000 members reside overseas. The Society is responsible for all aspects of the geological sciences including professional matters. The Society has its own publishing house, which produces the Society's international journals, books and maps, and which acts as the European distributor for publications of the American Association of Petroleum Geologists, SEPM and the Geological Society of America.

Fellowship is open to those holding a recognized honours degree in geology or cognate subject and who have at least two years' relevant postgraduate experience, or who have not less than six years' relevant experience in geology or a cognate subject. A Fellow who has not less than five years' relevant postgraduate experience in the practice of geology may apply for validation and, subject to approval, may be able to use the designatory letters C Geol (Chartered Geologist).

Further information about the Society is available from the Membership Manager, The Geological Society, Burlington House, Piccadilly, London W1V 0JU, UK. The Society is a Registered Charity, No. 210161.

Published by The Geological Society from:
The Geological Society Publishing House
Unit 7, Brassmill Enterprise Centre
Brassmill Lane
Bath BA1 3JN
UK
(*Orders*: Tel. 01225 445046
Fax 01225 442836)

First published 1995

The publishers make no representation, express or implied, with regard to the accuracy of the information contained in this book and cannot accept any legal responsibility for any errors or omissions that may be made.

© The Geological Society 1995. All rights reserved. No reproduction, copy or transmission of this publication may be made without written permission. No paragraph of this publication may be reproduced, copied or transmitted save with the provisions of the Copyright Licensing Agency, 90 Tottenham Court Road, London W1P 9HE. Users registered with the Copyright Clearance Center, 27 Congress Street, Salem, MA 01970, USA: the item-fee code for this publication is 0305-8719/96/$7.00.

British Library Cataloguing in Publication Data

A catalogue record for this book is available from the British Library.

ISBN 1-897799-51-9

Printed by City Print (Milton Keynes) Ltd, Milton Keynes, UK

Distributors

USA
AAPG Bookstore
PO Box 979
Tulsa, OK 74101-0979
USA
(*Orders*: Tel. (918) 584-2555
Fax (918) 560-2632)

Australia
Australian Mineral Foundation
63 Conyngham Street
Glenside
South Australia 5065
Australia
(*Orders*: Tel. (08) 379-0444
Fax (08) 379-4634)

India
Affiliated East-West Press PVT Ltd
G-1/16 Ansari Road
New Delhi 110 002
India
(*Orders:* Tel. (11) 327-9113
Fax (11) 326-0538)

Japan
Kanda Book Trading Co.
Tanikawa Building
3-2 Kanda Surugadai
Chiyoda-Ku
Tokyo 101
Japan
(*Orders*: Tel. (03) 3255-3497
Fax (03) 3255-3495)

Contents

Preface	iii
Introduction	v
1. Molten Earth - An introduction to igneous rocks and crystallization	1
2. 'All that glisters . . .' - An introduction to minerals	9
3. Whether it weathers - Processes of weathering	17
4. Shake, rattle and roll - Erosion	29
5. Sticking together - Sedimentary rocks	37
6. Feeling the strain - Metamorphism	47
7. Re-cycling the crust - The Rock Cycle	55
Appendix A - Concept development	67
Appendix B - Technician's guide	69
Appendix C - Resources	71
Appendix D - Glossary	75
Appendix E - An introduction to plate tectonics	77

Preface to the Second Edition

This handbook was originally written to meet the requirements of the 1989 Science Order at Key Stage 3. The 1995 edition of the handbook has been extensively revised and shows how the Earth science content of the 1995 Science Order can be delivered effectively as an integral part of 'Materials and their Properties'.

The handbook contains a series of classroom activities that have been devised to assist teachers in implementing the Earth science component of the Science Order at Key Stage 3. The topics and activities are designed to form an integral part of a process-science programme for the years 7 to 9. They require no prior knowledge of Earth science theory or nomenclature on the part of the pupil *or* the teacher. Each unit provides preparatory material for the teacher, the necessary information sheets for the pupils (where appropriate), the outline of a pupil investigation, and a homework project.

Progression

Royal Holloway Earth Science activities are intended to cater for the needs of children of all levels of achievement. Particular attention has therefore been given in this handbook to progression from Key Stage 2 to Key Stage 3, and from Key Stage 3 to Key Stage 4. With this in mind, the project team has been careful to include only that material which is at Key Stage 3 to ensure a coherent flow between the Key Stages.

Safety

We have attempted to check that all recognized hazards have been identified and suitable precautions are suggested. However, teachers should be aware that errors and omissions can be made, and that different employers adopt different standards. Therefore before doing any practical activity, teachers should always carry out their own risk assessment. In particular, any local rules issued by their employer must be obeyed, whatever is recommended here. We have assumed practical work will be carried out in properly equipped and maintained laboratories; field work takes account of any guidelines issued by the employer; care is taken with normal laboratory operations such as heating substances; good laboratory practice is observed when chemicals are handled; and eye protection is worn whenever there is any recognized risk to the eyes.

Department of Geology,
Royal Holloway, University of London
Egham,
July 1995

Introduction

The purpose of this handbook is to help the busy science teacher with no special expertise or training in Earth science to deliver the 'Earth science component' of the Science Order of the National Curriculum with the minimum of specialized equipment. The book provides a 'survival pack' of pupil activities that the teacher can use immediately in its present form, or could develop further if time allows. Indeed, the authors hope that teachers *will* use the handbook as a starting point for further development within their own schools.

The activities described in the handbook are designed to provide a course which forms a coherent whole and meets the requirements of the Key Stage 3 programme of study. Though the activities have been designed to cover what pupils should be taught through 'Materials and their Properties', the requirements of 'Experimental and Investigative Science' form a central part of each unit, so that the course as a whole can integrate readily into a process-based approach to science education. The activities can be used together to form an 'Earth science module', or can be drawn upon individually to suit a school's scheme of work.

For the purpose of teaching in schools, *copyright is waived for the reproduction of the Student Sheets and the Technician's Guide only.* All other material is copyright.

The format of the handbook was developed in consultation with a large number of practising science teachers. Each self-contained unit begins with a summary of the aspects of the National Curriculum to which the activity concerned refers, followed by an estimate of the time required for the average pupil to complete the work, and a list of the equipment pupils will require. The main concepts developed in each unit are included in Appendix A. A combined list of the equipment needs for all of the activities is given in Appendix B, a copy of which can be given to your laboratory technician to keep on file. The other appendices include sources of further information, suggestions as to what rock specimens you will need and where to obtain them, and a list of useful contacts and their addresses. Also included is a short glossary of terms referred to in the body of the handbook (only the bare minimum of specialized jargon has been introduced). To help the teacher see each unit in context, a summary of the Science Key Stage 3 programme of study having Earth science content is shown on the following pages as well as the relevant parts of the Geography Key Stage 3 programme of study.

Acknowledgements

Royal Holloway Earth Science was established in 1989 with the support of a grant from the Universities Funding Council. We thank Jenny Frost (University of London Institute of Education) for her contributions to the initial development of the project.

The preparation of the units in this book involved close liaison with teachers, and we thank them for their time and energy. We are particularly grateful to Chris King (Earth Science Teachers' Association), Dee Edwards (Open University) and Mike Brooks (Geological Society Education Offilcer). We thank Ready Mixed Concrete (UK) Ltd. for providing the quarry picture shown on p.45.

We thank the Association for Science Education Publications Committee for its recognition of the educational aspects of the activities contained within this publication.

Material from the National Curriculum is Crown copyright and is reproduced by permission of the Controller of HMSO.

Cover photography courtsey of Aznive Simons.

Summary of Earth science content at Key Stage 3

Materials and their properties

2. Changing materials

geological changes

2f how rocks are weathered by expansion and contraction and by the freezing of water

2g that the rock cycle involves sedimentary, metamorphic and igneous processes that take place over different timescales

2h that rocks are classified as sedimentary, metamorphic or igneous on the basis of their processes of formation, and that these processes affect their texture and the minerals they contain

chemical reactions

2n about chemical reactions that are generally not useful

2p about possible effects of burning fossil fuels on the environment

3. Patterns of behaviour

acids and bases

3i how acids in the atmosphere can lead to corrosion of metal and chemical weathering of rock

Physical processes

5. Energy resources and energy transfer

conservation of energy

5f that energy can be transferred and stored

The relevant parts of the Geography orders are reproduced below and referenced to the Science Orders.

Science Order	Geography Order
Key stage 2 **Sc3/2g rock cycle**	**Rivers;** how rivers **erode, transport** and **deposit** materials.
Key stage 3	**Thematic studies**
Sc3/2g rock cycle **Sc3/2h igneous rocks**	**7. Tectonic processes** In studying earthquakes or **volcanoes** and their effects on people, pupils should be taught: **a** the global distribution of earthquakes and **volcanoes** and their relationship with the boundaries of crustal plates; *and either* **b** the nature, causes and effects of earthquakes **c** about human responses to the earthquake hazard; *or* **d** the nature, causes and effects of **volcanic eruptions;** **e** about human responses to the volcanic hazard.
Sc3/2f weathering **Sc3/2g rock cycle** **Sc3/2h sedimentary rocks**	**8. Geomorphological processes** In studying geomorphological processes and their effects on landscapes and people, pupils should be taught: *either* **a** about the landforms associated with river channels, river valleys and drainage basins and the processes that form them, and about the **role of rock type** and **weathering** in landform development; **b** the causes and effects of river floods and how people respond to and seek to control the flood hazard; *or* **c** about coastal landforms and the processes that form them, and about the **role of rock type** and **weathering** in landform development; **d** the causes and effects of either cliff collapse or coastal flooding and how people respond to and seek to control the hazard.

1

Molten Earth

An introduction to igneous rocks and crystallization

1. Molten Earth - An introduction to igneous rocks and crystallization

This activity introduces pupils to the processes of igneous rock formation and the effects of volcanism on the Earth's surface.

National Curriculum teaching points

Sc3/2. Changing materials geological changes

2g that the rock cycle involves sedimentary, metamorphic and **igneous processes** that take place over different timescales

2h that rocks are classified as sedimentary, metamorphic or **igneous** on the basis of their processes of formation, and that these processes affect the texture and the minerals they contain

Practical details

Time required

This piece of work is intended to take one double period (i.e. 60-80 minutes) and a homework.

Materials needed for practical work

- *Beaker* - 250cm^3
- *Boiling tube*
- *Gauze*
- *Bunsen burner*
- *Tripod*
- *Safety mat*
- *Microscope slides*
- *Salol* (phenyl salicylate or phenyl-2-hydroxybenzoate); avoid contact with skin.
- *Safety goggles* - essential
- *Igneous rocks* - two types are needed showing a clear contrast in crystal size. In one the crystals should be clearly visible; **granite** is an obvious choice. The other should have much smaller crystals which are barely distinguishable to the naked eye, such as **basalt**. Appendix C tells you how to acquire suitable rocks.
- *Access to a refrigerator*
- *A magnifier or hand lens*

Understanding the Earth science

Igneous rocks are formed by the crystallization of cooling *magma* (molten rock) at temperatures between 800 and 1200°C. As the magma cools, crystals grow from the melt and interlock with each other, forming a hard crystalline rock.

In terms of mode of occurrence and appearance, igneous rocks fall into two classes:

Extrusive rocks

Extrusive igneous rocks are the result of magma being erupted at the surface, though a *vent* (pipe) or *fissure* (crack) to form lava. The magma may flow quietly on to the surface as a viscous liquid, producing a *lava flow*, or if it contains a lot of dissolved gases (mainly water vapour) it may be blown out by the explosive decompression of these gases. A useful analogy is the 'de-corking' of a pop bottle or can. Magmas that are low in volatiles erupt as liquid lava flows and can flow freely, whereas those that contain a high percentage of volatiles may produce explosive eruptions. Lava erupted at the surface cools rapidly, and solidifies as a mass of fine crystals (less than 1 mm in size), as in *basalt*. The reason for this is that the initial *nucleation* of crystals depends on the degree of supercooling to which the liquid is subjected: rapid cooling produces strong supercooling that leads to efficient nucleation. The result is a large number of nuclei that have little time or space to grow into large crystals. (In extreme cases however the lava may be cooled so quickly that growth of nuclei is suppressed, resulting in a volcanic glass [e.g. *obsidian*].)

Intrusive rocks

These are igneous rocks that have crystallized within the crust of the Earth. The volume of a large igneous intrusion may amount to tens, hundreds or even thousands of cubic kilometres. Because of the heat capacity of a large magma body, its large volume in relation to its surface area, and the insulating effect of the surrounding

1. Molten Earth - An introduction to igneous rocks and crystallization

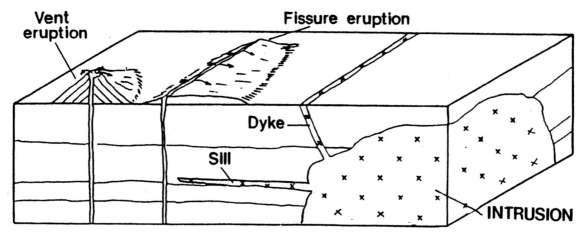

Figure 1.1 *A diagrammatic representation of the modes of occurrence of igneous rocks.*

crustal rocks, the cooling rate is slow, few nuclei form and large crystals are able to develop (5 - 25 mm). The result is a coarse-grained igneous rock such as *granite* (which is rich in silica [silicon (IV) oxide]) or *gabbro* (which is rich in iron-magnesium minerals).

You do not need to become fully familiar with the terminology of igneous rocks. The rock types mentioned in the text (granite, gabbro, basalt and obsidian) are among the most common types and are well suited for study by pupils. A flow chart showing a simple way of identifying rock samples in general is given on p.**73.** Table 1 (p.**6**) shows how igneous rocks may be distinguished.

Figure 1.1 shows several forms that igneous bodies may take. Intrusive igneous rock features seen at the Earth's surface today formed underground. A large intrusion takes millions of years to cool and solidify. Cornwall and Devon are underlain by a large intrusion of granite which extends from Dartmoor to the Scilly Isles. The dyke and the sill would cool much faster. The Romans used the Whin Sill in Northumberland as a platform on which to build Hadrian's Wall as well as a source of stone for the wall. An example of a vent eruption, forming a volcanic cone, is Mount Fuji in Japan. Fissure eruptions produce lava plateaus as seen in Iceland.

Teaching guidance

In this activity pupils experiment with the crystallization of salol (melting point 45°C) from the molten state as an analogue of the crystallization of a silicate melt (melting point 800-1200°C) and observing the effect of different cooling rates on the crystal size obtained. The activity can be enhanced if introduced by a few colour prints, slides or video (e.g. The Rocks Endure, Shell Education Service) illustrating volcanic activity and the rocks it produces.

The water-bath should be a few degrees above the melting point of Salol (45°C). The microscope slides must be completely dry.

You may, if you wish to save time, introduce the salol crystallization experiment as a demonstration. It is possible to watch the crystals growing using an OHP. (The glass slide should be supported on wood or card spacers above the surface of the OHP so that the salol is able to cool.) The advantage of projecting on to a screen is that the pupils will be able to distinguish clearly the form of the salol 'crystals', which actually consist of bundles of individual crystals radiating from a common nucleus[1]. Each bundle grows outwards until its growth is impeded by its neighbours. The slower the cooling rate, the fewer the nuclei that will form and the larger each bundle ('crystal') will be.

Using the rock specimens provided, the pupils should notice the different crystal sizes and suggest what factors may have caused these differences. There are several possible variables, one of which is the cooling rate. The pupils should then construct a fair test to investigate this, e.g., cooling equal amounts of salol, with the same initial temperature, at different rates.

---

[1] *This radiating structure only occurs rarely in real igneous rocks (it is usually associated with devitrification of volcanic glass), but the pupils do not need to be told of this difference. One or two more able pupils may notice it for themselves.*

1. Molten Earth - An introduction to igneous rocks and crystallization

Further development

A fruitful development is for the teacher to make up a supersaturated solution of copper sulphate and alum and leave it to evaporate for several days. Drain off the excess solution and view the crystal mass from the base of the beaker. The blue and white crystals grown in this way will interlock and appear very much like a typical piece of granite, although of course the colours differ from natural minerals. This experiment demonstrates that an igneous rock consists of a *mixture* of crystalline minerals, all of which have crystallized together from the same homogeneous melt (or, in this case, solution).

Answer to question on the Student Sheet

The specimen with large crystals will have formed deep in the crust, because here the magma can only cool slowly (because the insulating effect of the surrounding crust imposes a slower cooling rate), giving the crystals a large period of time to grow. The specimen with small or barely-visible crystals formed on the Earth's surface where extruded magma, known as lava, cools quickly.

Answer to questions on the Homework Sheet

1. An earthquake just prior to the eruption.

2. Over-run by lava, collapse due to weight of ash on roof, fire due to hot gases/lava/ash.

3. By pumping sea water on to the lava to cool it and accelerate solidification. In this way, engineers used the lava itself as a barrier to retain the molten lava behind.

4. The lava improved the port by creating a new sea defence. The hot lava was used as an energy source: water was passed through pipes in the lava and used to heat homes. In fact, volcanic ash also forms an excellent building material (strong but lightweight) and often improves the fertility of soil. The pupils should not be expected to know all of these; one would suffice.

1. Molten Earth - An introduction to igneous rocks and crystallization

Homework Project

Heimaey: A town under threat

Heimaey is a town on a small island just south of Iceland.

At two o'clock in the morning on 23rd January, 1973, the ground shook violently. The town clock stopped. A few minutes later, a fissure (crack) in the ground, 2 km long, burst open on the nearby volcano Helgafjell and lava poured out, advancing slowly but surely toward the little town's harbour and threatening to cut it off from the sea. Explosions also blasted tonnes of lava into the air, where it cooled quickly, falling to the ground to form a layer of ash that in places was over 2 m thick. Hot sulphurous gases and steam belched from the ground.

Heimaey is an important fishing port and the government of Iceland was determined to prevent the harbour being destroyed. Teams of engineers battled for more than nine weeks, until finally the lava flows were stopped and the harbour was saved. Though more than half of the houses in the town had been destroyed, the people were able to move back and start rebuilding their community.

Questions

1. What warning did the people of Heimaey receive before the volcano erupted?
2. In what ways do you think the houses might have been destroyed?
3. To prevent the hot molten lava flowing across the harbour mouth, the engineers had to find a way to make it solidify before it reached the harbour. How would you have done this? (Hint: Heimaey is right by the sea.)
4. Living near any volcano can be dangerous but may also offer a number of benefits. Did the eruption of Helgafjell bring any advantages to the residents of Heimaey and the surrounding area?

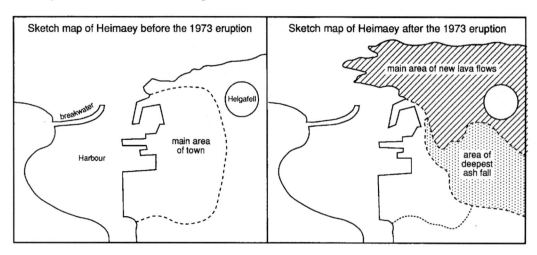

2 'All that glisters...'

An introduction to minerals

2. 'All that glisters...' - An introduction to minerals

Teaching guidance

The practical exercise
'ALL THAT GLISTERS...'

The investigation involves the pupil in identifying mineral specimens according to the properties listed in the Student Information Sheets 1 and 2. Each pupil should attempt to identify at least four different minerals. The choice will depend on availability, but the four should include *quartz* and *calcite* (very common rock-forming minerals that will be useful for subsequent exercises) and at least one mineral with a metallic lustre (e.g. *pyrite*, *galena*). For more information regarding the acquisition of specimens, teachers are referred to the Resources Section (Appendix C).

Note the following points:

a. It is advisable to test the steel nails against known minerals prior to the pupils using them. The hardness of steels can vary considerably.

b. A hardness test requires only a small scratch (2-4 mm). Advising the pupils of this can lengthen the life of the specimens considerably.

c. The hardness test should be performed on a clean, fresh surface. Weathered surfaces are likely to give low readings.

d. Encourage the pupils to use simple language when describing the density: does it 'feel heavy' or not?.

e. The acid reaction test only works reliably when the acid is applied to a clean, *dry* surface of the mineral. **Pupils should wear safety goggles when using dilute acid.**

f. Encourage the pupils to record other observations e.g. does the mineral shine like a metal?

If the school possesses sufficient good-quality specimens, it would be valuable for the other properties sometimes used in identification to be demonstrated. Alternatively, a series of activities could be set out around the laboratory, so that the pupils visit each in turn.

Industrial mineral separation (e.g. to separate ore minerals from worthless 'gangue' minerals) often exploits differences in mineral density. You may wish to exemplify this process by introducing a 'panning for gold' type of experiment.

2. 'All that glisters...' - An introduction to minerals

Student Investigation

'All that glisters . . .'

In the middle of the 1980s there appeared to be a gold rush in Botswana. Local people discovered a rich deposit of 'gold' in a local river. The news spread and soon thousands of people had packed their belongings and were on the move. They all expected to make their fortunes.

Unfortunately, they were all disappointed. The 'gold' was not really gold at all. It was a mineral that is sometimes known as "Fools' Gold". The proper name for this mineral, made up of iron and sulphur, is *pyrite*. A few simple tests could have shown up the mistake very quickly. Try some of these tests yourself to identify the mineral specimens provided.

❑ Select a mineral specimen and describe its hardness, crystal shape, density and reaction to acid. Use Student Information Sheet 1 to help you to do this.
❑ Record your observations in your notebook in a table like the one below.
❑ Use the information on Student Information Sheet 2 to help you identify your specimen.
❑ Repeat the tests for as many specimens as you have time for.

Specimen number	Hardness (Mohs' Scale)	Crystal shape	Density	Reaction to acid	Name
3	3	plates	heavy	none	barite

If you find a specimen that does not match any of those on the table, look in any good book on minerals. This will contain a lot more information.

3 Whether it weathers

Processes of weathering

3. Whether it weathers - Processes of weathering

2. *Chemical weathering* (or rock decomposition) is a very complex process, usually combining processes such as oxidation, carbonation, hydration and hydrolysis. The process modelled in Activity C is carbonation which particularly affects limestones (sedimentary rocks made up predominantly of the carbonate minerals, usually calcite, $CaCO_3$) and occurs as follows:

- rainwater dissolves carbon dioxide and becomes slightly acidic. [This acidity is accentuated by anthropogenic acid rain (H_2SO_4)]

$$H_2O + CO_2 \rightarrow H_2CO_3 \text{ (carbonic acid)}$$

- carbonic acid reacts with calcite

$$CaCO_3 + H_2CO_3 \rightarrow Ca(HCO_3)_2$$
(calcium hydrogen carbonate)

- calcium hydrogen carbonate is soluble and removed in solution.

Minerals in rocks react with the acids in the atmosphere, water and oxygen to form new substances. The products of chemical weathering are often softer and consist of fragments smaller than the original rock. Acids attack some minerals more easily than others. In the breakdown of granite, quartz [silicon (IV) oxide] often resists chemical weathering completely whereas feldspar and mica are commonly altered to form clay minerals.

3. *Biological weathering* Plant roots exploit weaknesses in rocks. As the roots grow they increase in girth and stress the rocks, enlarging the cracks and speeding up both mechanical and chemical weathering. The decay of vegetation produces acids (e.g. humic acid) which contribute to chemical weathering. In addition, organisms such as lichens slowly dissolve the rock they encrust and bacteria can aid the solution of silica.

Whilst the three types of weathering have been dealt with separately, it should be stressed that they rarely operate in isolation. Rather, they act in concert to break down rocks.

Teaching guidance

A useful starting point for this work is to ask pupils about why some buildings look older than others. Homework Sheet 1 can be used as an introduction to this unit.

Activity A
Investigating weathering by heat

It is desirable that the teacher selects the rocks to be used, and test them beforehand. Try to use distinctively different rock types, such as a soft sandstone and slate.

The rocks are heated indirectly as this is the safest method. Direct heating may generate internal stresses that cause the rock to shatter. For this reason, the activity should not be attempted with igneous rocks which may produce dangerously sharp fragments.

NB. Safety note — Safety Goggles must be worn during this activity.

Activity B
Investigating weathering by cold

The bottle must be completely full. If so, then the bottle should crack. Plastic 'mixer' bottles are ideal for this activity.

Activity C
Weathering by chemicals

One of the rock specimens treated must be a limestone (a sedimentary rock made up predominantly of the carbonate minerals, usually calcite, $CaCO_3$) and it is recommended that the other is granite. This will demonstrate that different rocks are affected in different ways by chemical action. In this case, the granite would not be affected at all. The rocks should be oven-dried prior to being placed in the carbonated mineral water.

There are two ways in which the teacher can organise these activities:

1. The pupils only do Activity A; Activities B and C are teacher demonstrations with the pupils noting the techniques used and the results and then applying the data in order to answer the questions.

3. Whether it weathers - Processes of weathering

2. The pupils initiate Activities B and C (it only takes about 10 minutes) and then leave them for a week before completing the exercise.

Further development

Experimental and Investigative Science for Key Stage 3 states that 'pupils should be taught to consider contexts, e.g. *fieldwork*, where variables cannot readily be controlled, and to consider how evidence may be collected in these contexts'. Before studying weathering in the laboratory, pupils could undertake fieldwork to see the effects of weathering. A sample homework sheet is included (Homework Sheet 1) which pupils could use as an introduction to the effects of weathering.

It is a simple observational homework on the effect of weathering on buildings and walls. Pupils are asked to observe and record the amount of weathering at the base of a building or wall and to compare it with the amount of weathering two metres from the ground. They should note down where the building or wall is situated and what it is made of and, if known, how old it is. One of the most obvious activities is to study the different amounts of weathering of different building materials in the same building (e.g. two different stones or stone and brick). The likely outcomes are that more weathering has taken place at the base of a building or wall; bricks become rounded on their edges and the cement gap deepens. Older buildings and walls show more signs of weathering. Buildings and walls close to roads will show more weathering. Ask the pupils to predict why this has happened. South facing walls tend to show a greater amount of weathering due to changes in temperature.

Fieldwork under teacher guidance can be done in any local graveyard, especially those where at least some of the stones have been standing for 50-100 years or more. Pupils can survey the gravestones and their observations can be used to stimulate discussion, perhaps leading to the pupils designing and carrying our their own experimental investigations into weathering.

There are several published strategies for graveyard visits, one of the best being 'Will My Gravestone Last?' (Science of the Earth, published by ESTA, available as a photocopy from Geo Supplies Ltd). Though this was written specifically for GCSE studies, it is easily adaptable for use with younger age groups.

If teachers do adopt the graveyard visit strategy, it is worth liaising with colleagues in other departments. It can form the basis of a good cross-curricular theme involving Science, English, Maths, Humanities and the Social Sciences.

Answers to questions on the Student Sheet

1. A simple answer may be 'make them break or crack' but more able pupils should include reference to expansion and contraction.

2. The natural process is much slower, as the range of temperatures is lower. The rocks are heated by solar radiation rather than immersion.

3. The volume increased. Able pupils may include reference to an 8% increase. The bottle split because the ice had a volume exceeding that of the bottle.

4. The water would freeze, expand and make the crack bigger.

5. The mass of the limestone was reduced, the granite remained constant. The carbonated mineral water partially dissolved the limestone.

6. Rain water would also dissolve the limestone, but much more slowly.

Answers to questions on Homework Sheet 1

See comments in **Further Development**, paragraph 2. This is an open-ended homework. Older buildings usually show more signs of weathering. Limestones and sandstones may weather more than brick.

Answers to questions on Homework Sheet 2

1. The average monthly temperatures are very similar. The mean monthly temperature for Timbuktu and Georgetown is 28°C to the nearest whole number. It is hot in both places!

2. The average monthly rainfalls are very different; Timbuktu is dry for most of the year whereas Georgetown has the most rainfall.

3. Whether it weathers - Processes of weathering

3. Timbuktu. Not chemical — the rainfall is too low. Not freeze/thaw, as there is insufficient water and temperatures rarely fall to freezing. Not by plants, as there is insufficient water. Therefore, thermal stressing is likely to be the most active.

3. Georgetown. High temperatures and rainfall would encourage plant growth and chemical weathering, of which the latter would dominate.

4. A bit of a trick, but the more able pupils should spot that very little weathering of any kind would operate here. However, an acceptable answer would be freeze/thaw.

3. Whether it weathers - Processes of weathering

Student Sheet

Whether it weathers

Old buildings, like churches and town halls, are often made of natural stone. They look very solid and strong and it might seem to us that they will stand for ever. This is not so. Over the course of time, rocks are broken down by processes that scientists call **weathering.**

Activity A - Investigating weathering by changes in temperature

1. Select two different types of rock, each about the size of a small egg.
2. Fill a beaker with water and heat it to about 100°C. Then place the rocks into the water and leave them for 3 minutes.
3. Fill another beaker with crushed ice. Using the tongs, remove the rocks from the water and place them into the ice. *Listen carefully.*
4. Leave the rocks in the ice for 3 minutes and then remove them. Study them carefully and note any changes.
5. Repeat steps 1-4 another three times.

Questions

1. What did heating and cooling the rocks do to them?
2. During the day rocks are heated by the sun and at night they cool down, so what you have done in this experiment happens naturally. How would this natural process differ from your experiment?

Activity B - Weathering by cold

1. Fill the measuring cylinder to the 100 cm³ mark.
2. Fill the bottle to the brim with water and screw on the cap.
3. Place the bottle in a polythene bag and close it with a tie.
4. Place both the measuring cylinder and the bottle in the freezer.
5. Leave them both until they are frozen.

Questions

1. What happens to water when it freezes? How did this affect the bottle?
2. Imagine that water has seeped into a crack in a rock. What might happen to the rock on a cold night?

Activity C - Weathering by chemical attack

1. Weigh small pieces of the two rocks provided.
2. Place each into a separate bottle of carbonated mineral water, and seal the bottle.
3. Leave the pieces of rock in the bottles for a week.
4. Remove the rocks, rinse with water, dry them thoroughly, and then reweigh them.

Questions

1. How did the weight of each rock change after it had been in the carbonated mineral water? Can you explain this?
2. Carbonated mineral water has a pH of 4. Rainwater has a pH of about 5.7. What effect would rainwater have on rocks like the ones in your experiment?

3. Whether it weathers - Processes of weathering

Homework Sheet 1

Buildings and weathering

Buildings are all around us. We expect them to last for a long time and to stand for ever. They must be weatherproof. How does the weather affect buildings around you? In this homework you are going to look at some buildings and see if they have weathered. You will need to record your observations so remember to take paper and pencil with you when you look at your building.

How to look for signs of weathering

Peeling paint, rusty metal and damaged bricks are signs of weathering.

Answer the questions below to help you decide if a building is being weathered:

1. How old is the building? Is there a date on the building e.g. Railway Terrace 1901?
2. Where is the building? Is it in a busy road or a quiet avenue?
3. What is the building made of? Is it completely made from brick or are parts of it stone?
4. Look at the base of the building: what has happened to the edges of bricks and the mortar?
5. Look at the same wall two metres from the ground: do you notice any differences in the edges of bricks or the mortar?
6. Have you seen any changes?

Write a short account about your building stating its age, where it is situated, what it is made from, where it is weathered and what you think has caused the weathering.

Extension work

Choose another building, statue or wall but this time look for a building, statue or wall that has both brick and stone in it. Answer questions 1 to 3 as before.

Look at the different materials that have been used. Which one has weathered the least? Which one has weathered the most? Remember to record where you observed the weathering: was it near the base, on a corner or somewhere else? Try and explain why one material is more resistant to weathering.

Write a short account of your findings. Remember to describe what you think has caused the weathering.

As you walk around your neighbourhood and your school, what signs and effects of weathering can you see?

3. Whether it weathers - Processes of weathering

Homework Sheet 2

Weathering around the World

The table below shows some climatic information for two different places. The world map shows you where these places are.

	TIMBUKTU		GEORGETOWN	
	Average monthly **temp.** (°C)	Average monthly **rainfall** (cm)	Average monthly **temp.** (°C)	Average monthly **rainfall** (cm)
January	21	0.5	27	20.0
February	23	0.4	27	12.5
March	27	0.3	28	17.5
April	32	0.1	29	13.5
May	35	0.5	28	28.0
June	33	2.0	27	30.0
July	30	7.5	27	25.0
August	29	8.0	27	17.0
September	32	4.0	30	9.0
October	30	1.0	29	8.0
November	26	1.0	28	12.5
December	22	0.5	27	28.0

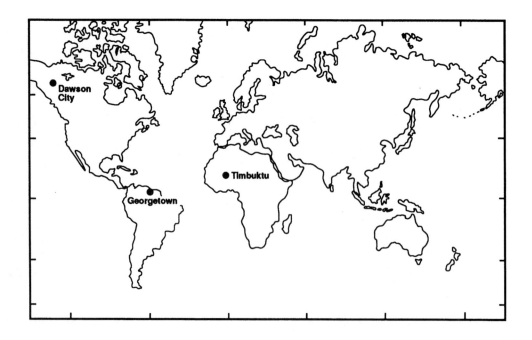

3. Whether it weathers - Processes of weathering

1. Compare the average monthly temperatures of Timbuktu and Georgetown: are they similar or different? Are the average monthly temperatures hot or cold?
2. Compare the average monthly rainfalls of Timbuktu and Georgetown: are they similar or very different? Which place has the most rainfall?
3. At which place is mechanical weathering more likely to take place? Explain your answer.
4. At which place is chemical weathering more likely to take place? Explain your answer.
5. The map also shows Dawson City in Canada. Here the temperature stays below freezing for at least seven months each year, and it is very dry. What types of weathering would you expect to take place in Dawson City?

4
Shake, rattle and roll
Erosion

4. Shake, rattle and roll - Erosion

This unit suggests a practical project in which students simulate the effects of erosion, and homework in which they consider how human activities may influence such processes.

National Curriculum teaching points

Sc3/2. Changing materials geological changes

2g that the rock cycle involves **sedimentary**, metamorphic and igneous processes that take place over different timescales.

Sc4/5. Energy resources and energy transfer conservation of energy

5f that energy can be transferred and stored

Practical details

Time required

This piece of work is intended to take one double period (i.e. 60-80 minutes) and a homework.

Materials needed for practical work

For each group:

- ❑ *20 pieces of brick small enough to fit in container*
- ❑ *Samples of crushed rock can be used as well as or instead of brick. Suitable examples are granite, sandstone and slate and are available from the suppliers listed in Appendix C - Resources*
- ❑ *Plastic container with wide neck*
- ❑ *Tray to collect fine residue when sieving*
- ❑ *Colander*
- ❑ *Access to a balance*
- ❑ *Graph paper*
- ❑ *Student sheet*
- ❑ *Homework sheet*

Understanding the Earth science

Erosion is an ever-continuing process driven by energy transfer. It is the physical removal ('transport') of rock fragments. Today's landscape is just a snapshot of a continually changing surface topography: surface rocks are constantly being weathered, eroded and re-deposited elsewhere. There are four principal agents of erosion:

Rivers

Fast rivers erode valleys. Slow rivers deposit flood plains. A high kinetic energy enables the river to erode solid material. If the energy drops, the river will deposit much of its load. The transport capacity of moving water is quite remarkable: a river with water moving at 24 km per hour (15 mph) is capable of moving boulders with a mass in excess of 1 tonne. The combined effect of the velocity of the river (which depends on the steepness of the terrain), its elevation above sea level, the volume of water flowing in it, the rock types over which it passes and their structure produces a wide variety of landscape features such as gorges, waterfalls and meanders.

Ice

During its long history, the Earth has experienced many periods when the climate was much colder than today, and widespread glaciation occurred at moderate latitudes. Much of Britain was covered by ice only 50,000 years ago. The icesheets scoured the landscape, wearing away the rocks and producing many characteristic glaciated landscapes, especially in the Highlands of NW Scotland, the Lake District and North Wales. Glacial features include U-shaped valleys, hanging valleys, moraines and drumlins. The material eroded was subsequently deposited as the glaciers retreated, forming the clay that mantles many parts of southern and eastern England.

Sea

Tides move enormous amounts of water. The energy of this moving water causes erosion in a similar way to that of rivers, but the rates involved are considerably greater. Many stretches of our own coastline are under threat from erosion, for example in the Peacehaven area of East Sussex, where the chalk cliffs are retreating at an average of 1 m per year.

Wind

In coastal and desert regions, wind is a most effective transporter of small particles which then 'sand blast' anything they collide with. In most other regions the wind is less effective as an agent of erosion, except where humans lend a hand by removing protective vegetation.

Teaching guidance

The process of erosion studied in this unit provides an opportunity for teachers to link this work with energy transfer, Sc4/5. Pupils need to understand that erosion occurs because of energy transfer. The student investigation 'Shake, rattle and roll' simulates an erosional process called *attrition*, which occurs when moving rock fragments collide with and grind against one another. It leads to a reduction in particle size.

Two student sheets are included. In Student Sheet 1, fragments of house-brick are shaken vigorously in a plastic container for a measured period of time, reproducing in a simple way what happens to rock fragments undergoing transportation. Brick is used because it is softer than many rocks and therefore shows measurable attrition on a timescale consistent with a classroom activity; it is also a reproducible material that is abundantly available at minimal cost.

The student experiment is the noisiest activity imaginable, so it would be advisable to warn colleagues in nearby rooms.

It is best if the pupils work in groups of 3-4; distributing the various jobs saves time. The recommended procedure is up to 5 shakes, each of one minute. After each minute of shaking, the contents of the plastic container are 'sieved' using the colander and the coarser fragments massed. The procedure can in principle be repeated until there are no pieces left in the colander, or until the pupils are exhausted, or until you can't stand the noise any longer!

Given sufficient time, the exercise can be extended by altering the variables as follows:

❑ add a little water
❑ use different types of rock fragment
❑ use a particular range of fragment size.

Student Sheet 2 is more open ended and could form the basis of an investigation for the **Experimental and Investigative Science Attainment Target (Sc1)** assessment. Experience shows that each school has its own way of approaching Sc1 investigations and you may wish to use this sheet as a basis for producing your own.

The homework exercise examines the wearing away of surfaces by transported materials, a process known as *abrasion*.

Answers to questions on the Student Sheet

1. They have become rounder and smaller.
2. The answers here will be very variable, but pupils must be able to critically analyse their own ideas.
3. In water (which acts as a lubricant) the process would have been slowed considerably.

Suggested mark scheme for 'Shake, rattle and roll' investigation

Key features

1. Planning experimental procedures

   level 3 - when you shake the rock breaks up.

   level 4 - the longer you shake the greater the wear; the longer you shake the rounder it gets; the longer you shake the smaller the rock fragments become.

   level 5 - the more energy available the greater the wear; different rocks wear at different rates because they are made up of different materials.

2. Obtaining evidence

   level 3 - to record that the rock broke into smaller pieces.

   level 4 - measurement of mass of rock at beginning and at end.

   level 5 - measurement of mass loss over 'time'.

4. Shake, rattle and roll - Erosion

3. Analysing evidence and drawing conclusions

   level 3 - when rocks are shaken, they break up.

   level 4 - the amount of mass lost depended upon the strength of the rock and the length of shaking.

   level 5 - the amount of wear depended upon the type of rock (e.g. hardness), the time it was shaken and whether it was shaken dry or wet.

4. Considering the strength of evidence

   level 3 - it was difficult to shake at the same amount each time but only one person shook so we think our results are fair.

   level 4 - we used the same number of rock pieces at the start because it makes the 'bashing' fair.

   level 5 - we used the same number of rock pieces at the start and tried to have exactly the same mass of rocks; when we used water we did not dry the samples completely when we weighed them, but the mass lost was always greater than the water; although our results were not completely accurate they did show a pattern.

Answers to questions on the Homework Sheet

1. The pebble from A has not undergone erosion; it has sharp edges and pointed corners. The B pebble has undergone river erosion. The edges and corners are slightly rounded. The pebble from C has been subjected to more intensive erosion in the tidal zone, where attrition by pebbles carried by tides and waves has rounded the fragments.

2. Inland: the roundness of the granite pebbles in relation to their position suggests an inland source.

3. The sandstone cliffs are being eroded by the hard granite pebbles being 'thrown' against them by the sea. This process is concentrated at the base of the cliff. This question can be used to emphasize that it is the **combination** of the moving water **and** the pebbles it carries that causes the erosion.

4. There are various options available here:

   (i) construction of protective barrier at the base of the cliff; possible materials include concrete (expensive, with engineering difficulties), large blocks of resistant rock (cheap if suitable material available locally), or resistant pebbles in retaining cages;

   (ii) grading of the cliff, i.e. reduction and smoothing of the cliff gradient;

   (iii) construction of breakwaters (or groynes) to prevent movement of the granite pebbles;

   (iv) construction of a breakwater out at sea. (Another technique used on parts of the South Coast today is to haul shingle by road and dump it at the 'up-current' end (in this case the western end) of the area prone to erosion. The current energy is then dissipated by moving the shingle and the erosion is greatly reduced.)

Student Sheet 1

Shake, rattle and roll

In this exercise, you will be studying what happens to pieces of rock when they are moved around and collide with each other. What you are going to do is to put 20 bits of brick into a container and shake them around. BUT before you start you must make a *prediction:* you must write down what you think will happen to the bits of brick.

1. Pick out 21 pieces of brick. Draw and describe the shape of *one* of the pieces; keep it aside for comparison at the end of the experiment.
2. Measure the total mass of the remaining 20 pieces of brick.
3. Put all of the pieces in the container and screw the lid on tightly.
4. Shake the container vigorously for one minute exactly.
5. Empty the container into the colander. Select all the pieces which are more than 5 mm across. Count these and weigh them.
6. Put all the pieces over 5 mm back into the container.
7. Repeat stages 3 to 6 as many times as you can.
8. When you have finished, draw and describe one of the pieces.
9. Plot your results in the form of a graph or graphs.

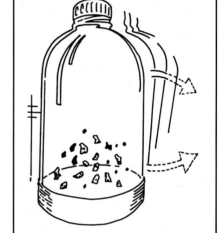

Questions

1. How has the shape of the pieces of brick changed?
2. How good was your prediction?
3. Of course in nature, large pieces of rock don't get shaken around in air, but they do get moved around in water. What difference do you think it would have made if you had put water in your container?

4. Shake, rattle and roll - Erosion

Student Sheet 2

Eroding rocks

The problem

Rocks are broken when they collide. Your task is to find out what factors affect the destruction of rocks.

Apparatus available

- *Plastic container and lid*
- *Tray*
- *Access to balance*
- *Pieces of rock* - granite, sandstone, slate.

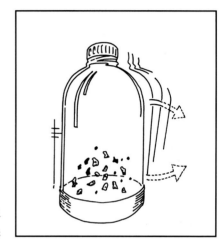

The problem

Which factors may affect how much rocks are worn when they are moved? Which of these factors are you going to vary and which are you going to keep the same? What background knowledge do you have that may help you? Before you start, make a *prediction:* write down what you think will happen to the rock pieces. Can you explain why?

Method

How will you do the experiment?

Will you need to make measurements?

What factors will you keep the same?

What factors will you change?

How will you present your results?

Is the experiment accurate and reliable?

Conclusion

Have your results helped you to find out about what things affect the way rocks wear? Is there a pattern in your results? If you investigated several things, which one had the most effect on wear and which the least? Can you explain your conclusion using scientific ideas and knowledge? Could you improve the experiment in any way?

4. Shake, rattle and roll - Erosion

Homework Sheet

The case of the disappearing cliffs

Tumbledown is a small village near the coast. It is built near the top of some cliffs. The cliffs are made of a soft sandstone, and are being worn away quickly by the sea. The local Council is very worried about this and has asked an engineering geologist to investigate the problem for them. The engineering geologist noted that the beach consisted mainly of pepples of granite, which could also be seen in the bed of the stream. In the report, the engineering geologist included the map and diagrams shown on this page. Use this information to answer the following questions:

1. The pebbles the engineer collected are all made of the rock, granite. This is a very hard rock. Can you explain why the pebbles from different places have different shapes?
2. Where do you think the granite pebbles may have come from? From inland, or from somewhere else along the coast?
3. Why do you think the sandstone cliffs are being worn away so quickly?
4. Suggest what the Tumbledown Council could do to stop (or slow down) the wearing-away of the cliffs.

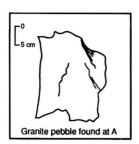
Granite pebble found at A

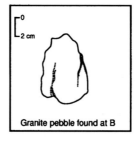
Granite pebble found at B

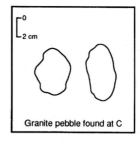
Granite pebble found at C

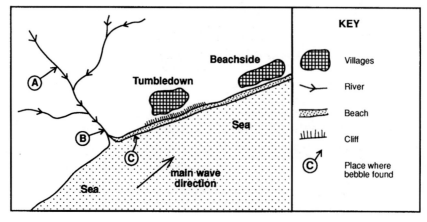

5 Sticking together

Sedimentary rocks

5. Sticking together - Sedimentary rocks

This unit introduces the student to the **transportation**[1] of the products of weathered materials in rivers and streams, and the **deposition** of sediment where a river flows into a lake or the sea. It demonstrates how the phenomenon of *bedding* is produced in sediments, and helps the pupil to appreciate the difference between unconsolidated *sediment* and a hard **sedimentary rock**. The environmental impact of limestone quarrying is included as an application of science.

National Curriculum teaching points

Sc3/2. Changing materials geological changes

2g that the rock cycle involves **sedimentary**, metamorphic and igneous processes that take place over different timescales

2h that rocks are classified as **sedimentary**, metamorphic or igneous on the basis of their processes of formation, and that these processes affect their texture and the minerals they contain.

Programme of study
2. Application of science

2d consider the benefits and drawbacks of scientific and technological developments in environmental and other contexts.

Practical details

Time required

This piece of work is intended to take one double lesson (i.e. 60-80 minutes) and associated homework.

[1] *Transportation and deposition are processes within the rock cycle, sediment and sedimentary rock are the resulting products.*

Materials needed for practical work:

- *Piece of sandstone*
- *Petroleum jelly*
- *Damp sand*
- *Powdered clay*
- *Plaster of Paris*
- *Plastic Syringe*

The sand should be fine in size, and only slightly damp. You may find it best to make up the mixtures for the pupils. Plastic syringes suitable for the activity are cheaply available from most suppliers. A cheap alternative to a plastic syringe is a piece of plastic tubing and a dowel rod. For obvious safety reasons *you should not use discarded medical syringes*.

Understanding the Earth science

A large part of the Earth's land surface is made up of **sedimentary rocks**. They consist of **sediment** (natural rock debris mostly derived from the *erosion* of the land surface) that has been cemented together to form a hard rock. Deposition of sediment can occur on land (terrestrial sediments), but most sedimentary rocks are the result of deposition under the sea (marine sediments).

The faster a river flows, the more sediment it can carry. Slowing a river down causes it to deposit some of its sediment load; this may occur where the terrain flattens out or where the river enters a larger body of water such as a lake or the sea. As the water velocity decreases, the coarser particles of sediment, because they sink faster, tend to be deposited first, while the finer material remains suspended longer and is carried further. This commonly results in the *sorting* of fine particles from coarse particles. Generally coarse sediments are deposited closer to their source, in near-shore environments, whereas fine sediments tend to be carried further out to sea.

5. Sticking together - Sedimentary rocks

Once deposited, the loose sediment usually undergoes various changes that convert it into hard rock. These changes, known to the geologist as *lithification*, involve:

Compaction: sediment is compressed by the weight of overlying layers of sediment deposited subsequently. This reduces the porosity of the sediment (the volume percentage of voids between grains).

Dewatering: compaction is accompanied by the expulsion of pore water, which in mud can constitute as much as 60% of the initial volume of the sediment.

Cementation: water continues to circulate through the compacted sediment and gradually precipitates salts between the grains of sediment.

These processes combine over long periods of time to convert loose sediment into hard sedimentary rock. Compaction and cementation are simulated in the classroom experiment associated with this unit.

The most obvious feature of many sedimentary rocks, when seen *in situ*, is that they consist of layers or *beds*. This *bedding* is noticeable because individual beds differ from each other in grain size, or in grain composition, or in resistance to erosion (i.e. how firmly the grains are cemented together). A succession of beds resting one upon another, as seen for example in a quarry face or cliff, represents a *time-sequence* of deposition, the lowest beds being the oldest.

Table 2 shows the main sub-divisions of sedimentary rocks and some common examples. See also the identification flow chart on p.**73**.

Table 2. Basic terminology of sedimentary rocks

Class of sedimentary rock	Rock names	What the name means	Particle size range (mm)
Clastic rocks Sedimentary rock consisting of mechanically transported particles of pre-existing rocks.	Conglomerate	A rock in which the larger particles ('clasts') are more than 2 mm in diameter.	>2.0
	Sandstone	Rock consisting of sand grains (mainly quartz). Sandstone can vary in colour owing to different kinds of cementing material.	0.06 - 2.0
	Mudstone	A fine-grained sedimentary rock.	<0.06
Biogenic rocks Sedimentary rocks consisting mainly of materials originally secreted or grown by organisms.	Limestone	Rock consisting essentially of carbonate minerals (mainly *calcite*, $CaCO_3$) secreted by marine or other organisms. Limestones commonly contain remains of these organisms in the form of fossils (shells etc.).	
	Coal	Rock consisting of coalified plant remains.	
Chemical rocks Sedimentary rocks consisting of chemically precipitated minerals.	Evaporite	Rock consisting of salts (minerals such as *halite*, NaCl, and *gypsum*, $CaSO_4.2H_2O$) precipitated from seawater by evaporation.	

Teaching guidance

Classroom experiment: Loose sediment into hard rock

The experiment introduces the pupil to some of the differences between unconsolidated *sediment* and hard sedimentary *rock*. The student tests the strength of simulated 'rock' pellets produced with various combinations of compaction and cementing.

The pupils should make up their own strength test. The possibilities include:

1. *Load-bearing capacity:* stacking 10-100 g masses onto the pellet until it collapses.

2. *Cohesiveness:* dropping the pellet from a measured height and measuring the resultant spread of the fragments.

3. *Hardness:* scratching the pellet with a copper coin and a piece of mild steel.

Note that only a *thin* smear of petroleum jelly is needed; its purpose is to facilitate removal of the pellets from the syringe. If too much is used, it may seriously affect the outcome of the experiment.

Homework experiment: Depositing sediment

1. The pupils should notice some degree of *sorting*, i.e. the larger fragments being concentrated at the base of each layer, the finer fragments at the top. (In geological terminology, each layer is said to be 'graded'.)

2. The experiment shows that coarse particles sink faster than finer ones.

3. Applying this general rule (basically Stokes' Law) to sediment suspended in a river running into the sea:

    (a) the coarsest particles will settle out soonest, near to the shore

    (b) the finer particles remain in suspension longer and are carried by currents further out to sea.

4. The oldest layer will be the one at the base of the cliff. The pupils are applying a fundamental principle of stratigraphy, the *Law of Superposition*, which states 'in an undeformed sequence of sedimentary strata, the oldest will be at the base of the sequence, the youngest at the top'.

Homework Study: Should we mine it?

Statements 2, 3, 7 and 9 could be used as arguments to support the quarrying proposals, whilst numbers 1, 4, 6 and 8 are negative effects. Item 5 could be interpreted either way.

5. Sticking together - Sedimentary rocks

Classroom Experiment

Loose sediment into hard rock

Compare the loose sand with the lump of sandstone. What's the difference between them? What makes the sandstone hard? A long time ago the sandstone was loose sand on the sea floor. Since then, the grains have become stuck together. How has this been brought about? You are going to explore some ways of converting sand into 'sandstone'.

1. Smear a *small* amount of petroleum jelly around the inside of your 'syringe'.
2. Fill the syringe with damp sand.
3. Put your finger over the open end of the syringe and press the plunger in as hard as you can.
4. Take your thumb away and gently push the sand pellet out of the syringe onto a piece of paper.
5. Repeat the procedure, this time using a mixture of damp sand and clay (3 parts sand:1 part clay).
6. Repeat the procedure again, but this time use a mixture of damp sand and plaster of Paris (5:1).

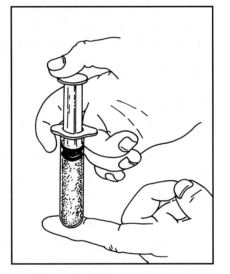

7. Leave your three pellets to dry for 10-15 minutes.
   While you are waiting, try to predict which pellet will be the strongest. Why?
8. Devise your own test to decide which is the strongest pellet, and then carry it out.
9. Do you think your experiment is a fair test? How would you explain the results?

Which factor had more effect in producing a hard, strong pellet? Applying pressure or adding 'Ingredient X'? How do real sediments get compressed? Is this enough to explain the hardness of the sandstone? If not, what else might have happened to make it hard?

5. Sticking together - Sedimentary rocks

An Experiment to do at home

Depositing sediment

Think of a swollen river after a long period of rain. It looks brown or milky because of the large amount of solid material (mud, sand and even small stones) being carried along in the water. The faster a river flows, the more solids it can transport.

Most of stones, sand and mud carried by rivers and streams will eventually reach a lake or the sea. When this happens, the moving water slows down (it 'loses energy'). Slower-moving water is less able to transport solid particles, which will begin to sink to the lake- or sea-floor, where they accumulate as *sediment*.

Why do sediments and sedimentary rocks form distinct layers? To find out, you will need a large, clear plastic container, some soil from the garden, and water.

1. Half fill your container with water.
2. Carefully sprinkle some soil into the container, enough to make a layer about 1 cm thick.
3. Leave the container in a safe place and allow the water to settle for two days.
4. Then add another layer and again leave it for two days.
5. Finally, add one more layer, allowing it to settle for two days.
6. Carefully record and measure what happens in the container *every* day.

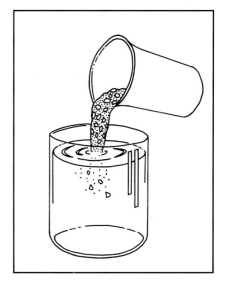

Questions

1. What happened to the pieces of soil in your experiment?
2. How does the *size* of the soil particles affect their behaviour?
3. Imagine the soil being carried as sediment in a fast-moving river as it flows into the sea? Which parts of the soil would be deposited (a) near to the shore, and (b) further out to sea?

5. Sticking together - Sedimentary rocks

4. The picture here shows a cliff face. It is made up of *beds* (layers) of sedimentary rock. Which layer is the oldest?

5. Sticking together - Sedimentary rocks

Homework Sheet

Should we mine it?

Limestone is a sedimentary rock made up mainly of a mineral called *calcite*. Calcite is very important in industry: it is used to make cement and glass, and is also used in the production of iron and steel. For a limestone to be worth extracting commercially from a quarry, it must be very pure (not have minerals other than calcite in it).

A mining company have discovered a new deposit of pure limestone which they want to quarry; they have applied to the local District Council to get planning permission for the quarry. Imagine you lived near to where the new quarry would be. How would it affect you? Read carefully the following statements.

1. Before quarrying can commence, the top soil must be removed.
2. The area has high unemployment and a large population of people in the 12-16 age-range.
3. Quarries have limited life spans. This one would be worked for eight years.
4. Quarrying is a noisy activity and also produces a lot of dust.
5. Limestone is bulky. It is not economical to transport it over great distances. This may encourage limestone users to move close to the quarry.
6. The quarry would have no rail access. Many large lorries would be needed to carry the limestone to the user.

5. Sticking together - Sedimentary rocks

7. We produce millions of tonnes of rubbish each year and need to dispose of it.
8. Water supplies can be affected by quarrying.
9. In areas where unemployment is *low*, local shops sell more goods.

Using this information, write two paragraphs. In the first, you should say what the *benefits* of the quarry would be, and who would gain most from it, and in the second you should estimate what *harm* it might do to the area.

If you were the local Planning Officer and had to advise the Council on the impact of the quarry, would you recommend them to authorize the quarry, or to turn it down? Give reasons for your answer.

6 Feeling the strain

Metamorphism

6. Feeling the strain - Metamorphism

This investigation simulates what happens when a rock is deformed. Equi-dimensional marks are applied to the surface of a block of playdough which is then squeezed by hand as an analogue of the deformation of rocks. The homework project exploits the familiar properties of slate (a metamorphic rock) to illustrate how different types of rock lend themselves to different uses.

National Curriculum teaching points

Sc3/2. Changing materials geological changes

2g that the rock cycle involves sedimentary, **metamorphic** and igneous processes that take place over different timescales

2h that rocks are classified as sedimentary, **metamorphic** and igneous on the basis of their processes of formation, and that these processes affect their texture and the minerals they contain

Practical details

Time required

This piece of work is intended to take one double lesson (i.e. 60-80 minutes) and associated homework.

Materials needed for practical work

❑ *Playdough* - playdough is a cheap, clean alternative to other materials. The ingredients are:

> Flour - 2 cups
> Water - 2 cups
> Salt - 1 cup
> Oil - 2 tablespoons
> Cream of Tartar - 2 teaspoons
> Colouring as required

Mix the ingredients together and heat over a low flame. Stir continuously until a ball forms. Allow the mixture to cool and then knead. This recipe makes enough playdough for 4-5 groups of pupils.

Playdough has the advantage over Plasticine of being much cleaner to use and is more pliable. It can be safely stored for up to six months if kept in an airtight container.

Understanding the Earth science

Metamorphic rocks are formed when pre-existing rocks are modified in the solid state by the effects of heat and/or pressure. The name is derived from the Greek, *meta* meaning 'change' and *morphe* meaning 'form'.

Metamorphism in its broadest sense can be seen as a combination of two processes:

(i) Re-crystallization. Rocks formed at the surface of the Earth may undergo burial through prolonged deposition of sediments on top of them, or as a result of mountain-building processes. They find themselves much deeper in the crust, where pressure and temperature are much greater than they are near the surface. Some of the original minerals in a rock may be unstable under the new conditions, and will react or re-crystallize into new minerals having crystal structures more compatible with high temperature and pressure. The process, which often involves the expulsion of water from hydrous minerals, is akin to the firing of china clay to produce porcelain.

(ii) Deformation. As well as experiencing an increase in hydrostatic pressure and temperature, rocks caught up in mountain-building episodes in regions like the Himalayas are also subjected to enormous lateral stresses. Most metamorphic rocks have been formed in a non-hydrostatic stress regime (i.e. the stress is greater in one direction than the others). Accordingly the rocks are compressed more strongly in that direction. This 'flattening' causes the growth of platy minerals (such as mica) perpendicular to the direction of maximum stress. Slate is again a good example: it can be split into thin sheets because the platy crystals it contains have crystallized parallel to each other as a result of the stress to which the rock was subjected during re-crystallization.

'Regional' versus 'Contact' Metamorphism

The metamorphism (re-crystallization and deformation) that occurs during mountain-building affects the whole region in which the mountain belt is developing, and is accordingly referred to as *regional metamorphism*.

Metamorphism on a more restricted scale occurs when an igneous intrusion is emplaced into the crust. The magma will initially have a temperature of 800-1200°C, and the 'country rocks' in contact with it will be heated up by thermal conduction. This baking produces a zone, called an *aureole*, of metamorphosed country rocks that have re-crystallized at high temperature but may have experienced little deformation. This phenomenon is called *contact* or *thermal metamorphism*.

It is not necessary to be familiar with the nomenclature of metamorphic rocks at this level but, in case you have to deal with questions from students, here are some simple definitions (see also the identification flow sheet on p.**73**):

Slate — The most familiar metamorphic rock, usually dark grey or black in colour. Mineral particles are *too fine to be seen with the naked eye*. Useful because it cleaves readily into thin plates (reflecting the parallelism of mineral grains on the fine scale). Formed by metamorphism of fine-grained sedimentary rock (e.g. mudstone) or volcanic ash.

Schist — Any moderately homogeneous metamorphic rock that exhibits *parallelism of platy mineral particles* (typically mica). Crystals are coarse enough to be clearly visible to the naked eye. Formed from sedimentary rocks such as mudstones, subjected to higher temperatures than slate. (The activity illustrates how the parallelism develops.)

Gneiss — A coarse-grained metamorphic rock that shows distinct banding (alternating layers of different composition). Such rocks are commonly the result of the most intense metamorphism, and can form from virtually any precursor: sandstone, shale, schist, even granite.

Marble — Geologists reserve this term for a *metamorphosed limestone*; decorative versions often have swirly patterns whose colours are due to silicate impurities resulting from metamorphism.

Caution. Stonemasons tend to use 'marble' for any decorative rock, usually limestone, that takes a polish, whether metamorphosed or not, e.g. 'Black Marble' is polished basalt, an igneous rock!

Teaching guidance

The processes of metamorphism are difficult concepts. The enormous forces involved and the increasing temperature with depth within the Earth are beyond our immediate experience. The student activity introduces the changes caused by metamorphism at the simplest level: what happens to a rock when it undergoes ductile deformation. Equi-dimensional marks are applied to the surface of a piece of playdough, which is then deformed by hand or by compression in a vice. Deformation is a more obvious consequence of the forces involved than the mineral re-crystallization, and it is more amenable to simulation and investigation in the classroom.

Geologists investigating deformed rocks look for similar small-scale structures (e.g. recognizable fossils in deformed sedimentary rocks) that are likely originally to have been equi-dimensional; measuring *how much* they have been flattened from their original shape provides an estimate of the degree of deformation (strain) that the rock as a whole has experienced.

During regional metamorphism the particles within a mudstone are re-crystallized and realigned to produce a slate. This activity simulates the process in a simplified way. You could allow the pupils to make their own playdough, as this models a metamorphic process in that the ingredients are changed from their former state by the effect of heat.

There should be discussion of the results of the activity prior to the pupils answering the questions. What should be seen is that the indentations develop a parallelism. At this point it would be advantageous for the pupils to have access to specimens of mudstone and slate. Old roofing slates can be broken up to provide excellent teaching specimens.

6. Feeling the strain - Metamorphism

The 're-crystallization' aspect of metamorphism can be demonstrated by the teacher using playdough into which copper sulphate crystals have been introduced. Make sure the crystals are readily visible on the surface of the block. Leave the block to dry out for 2-3 hours, and then place it in an oven or kiln at 180°C. Leave it for an hour and then remove. The blue copper sulphate crystals will have turned white, thus simulating the formation of new minerals in metamorphism. The copper sulphate will have changed from blue $CuSO_4.5H_2O$ crystals to white $CuSO_4$ crystals due to the expulsion of H_2O from the crystal structure as the temperature increased. This simple laboratory reaction mimics similar 'dehydration' reactions which minerals like clay and mica undergo during metamorphism.

Answers to questions on the Student Sheet - All change

1. The particles would become approximately parallel to one another and the pore space would be reduced.

2. The particles that make up the slate have been squashed to be parallel to each other. This arrangement means that the rock is easily split in this parallel direction.

3. The mudstone is the 'parent' of the slate.

Answers to questions on the Homework Sheet - Using rocks

1. The reasoning is the important aspect here, so accept any well reasoned answer. However, the most likely are:

    (a) limestone, as it is soft and easily shaped.

    (b) granite, attractive and durable.

    (c) slate, hard and strong, and non-porous, plus it splits easily and therefore would reduce weight.

    (d) pumice, which is lightweight but strong.

2. Possible reasons include:

    (a) cost

    (b) availability

    (c) transport problems

    (d) bricks are a fixed shape and are easy to handle / quick to put together

    (e) suitability

    (f) exhaustion of supplies of natural rocks

Obviously, these all interlink.

6. Feeling the strain - Metamorphism

Student Sheet

All change

Rocks may seem very strong and everlasting to us, but they can be changed from one type to another. This activity will show you how this could happen.

1. Make a block of playdough about 8 cm x 8 cm x 4 cm.

2. Using the lid of a biro or a pencil, make a random pattern of indentations on one surface of the playdough.

3. Gently compress your playdough block as shown in the diagram and note what happens to the indentations you made.

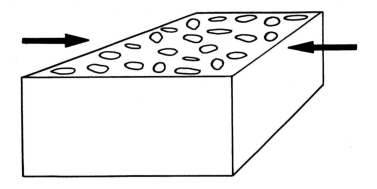

Questions

1. The particles that make up the rock mudstone are small (less than 0.0625mm across) and are randomly arranged. There are many gaps, or pores, between them. What would happen to the particles and the gaps in the mudstone if it was put under tremendous pressure?

2. Look at the piece of slate provided. You should notice that it splits easily. Can you explain why?

3. What is the relationship between slate and mudstone?

6. Feeling the strain - Metamorphism

Homework Sheet

Using rocks

Use the information in the table below to help you answer the questions.

Rock Name	Properties
Marble	White/green/red in colour often banded, takes a polish, quite strong, soluble, non-porous
Granite	Hard, very strong, insoluble, non-porous, an attractive speckled rock that takes a polish
Slate	Dark grey/black in colour, hard, strong, insoluble, non-permeable, can be cleaved into thin sheets
Pumice	Soft, strong, insoluble, porous, very lightweight (not found in the UK)
Limestone	Often light grey or brownish yellow, dull surface, permeable, soluble, easily shaped

Explanation of terms

1. *Hard* means the rock cannot be easily scratched.
2. *Strong* means that the rock can support a lot of weight before crumbling.
3. *Soluble* means that the rock can be slowly dissolved by water.
4. *Porous* means that the rock has pore spaces in it and will hold water.
5. *Permeable* means that water can flow through the rock.

Questions

1. Which rock would you use for the following purposes:

   (a) to make a statue

   (b) as a gravestone

   (c) as a roofing material

   (d) to make breeze blocks, which are made by mixing fragments of rock with cement.

   Remember to give the reasons for your answers.

2. These days we rarely use rocks to build houses. Can you suggest reasons to explain why we now use man-made materials like bricks, breeze-blocks and concrete rather than rocks?

7 Recycling the crust

The Rock Cycle

7. Recycling the crust - The Rock Cycle

This unit invites pupils to apply the understanding that they have derived from earlier units to the problem of classifying rocks into igneous, sedimentary and metamorphic categories. It helps them to appreciate the diverse environments in which each type of rock is formed, and that rocks may be recycled to form new rocks.

National Curriculum teaching points

Sc3/2. Changing materials geological changes

2g that the rock cycle involves sedimentary, metamorphic and igneous processes that take place over different timescales

Practical details

Time required

This piece of work is intended to take one double lesson (i.e. 60-80 minutes) and associated homework.

Materials needed

The pupils will need access to 5-6 rock specimens per group, including at least one igneous (i), one sedimentary (s) and one metamorphic (m) specimen (to facilitate assessment, each should be marked with a unique index letter or number). A suitable list would be:

granite (i)   basalt (i)   slate (m)
limestone (s) sandstone (s) mudstone (s)

Pupils will find it helpful to have the use of a hand-lens or magnifier in describing the specimens. Pupils might ask to test the rocks with dilute hydrochloric acid; the usually safety precautions of wearing safety goggles must be adhered to. Also the rocks will need washing to remove the acid! It is simpler and safer if pupils are told which rocks react with hydrochloric acid.

Understanding the Earth science

The Earth's crust is made up of all three major categories of rock whose formation has been considered in foregoing units. The Earth is an intensely active planet whose crust is subject to constant change. Surface rocks are continuously being weathered and eroded, their debris is transported, deposited as sediment and lithified to form new sedimentary rock. Rocks of all kinds, caught up in the deeper parts of mountain belts (for example, along the compressional zone between two colliding continents), undergo heating and compression and form new, metamorphic rocks. As erosion removes the mountains above over tens or hundreds of millions of years, these metamorphic rocks may eventually be exposed at the surface, where they will be subject to weathering and erosion and thereby provide the material for new sedimentary rocks. Magma continues to ascend from the mantle or the deep crust to form new igneous rocks. In trying to understand how the Earth works, we must recognize that the three major classes into which we divide rocks are immutable. But a sedimentary rock may not remain a sedimentary rock for ever; it may be transformed in due course into a slate, or even into a granite. This continual re-cycling of crustal materials has been going on since the Earth was formed; we refer to it as the *Rock Cycle*. The rock cycle is driven by plate tectonic processes. Appendix E contains supplementary information about plate tectonics. Whilst this information relates to Key Stage 4, it aids teacher understanding of the mechanism that drives the rock cycle. It is not intended that pupils should study plate tectonics at Key Stage 3.

Teaching guidance

This unit draws together in a summary the processes and products that the pupils have studied in previous units. The rock cycle involves the following processes:

❑ **weathering** - rock is broken down into smaller fragments

❑ **erosion/transport** - running water, wind or ice carry rock fragments into environments where they are deposited to form new rocks. (KS2 Geography - rivers)

7. Recycling the crust - The Rock Cycle

- **deposition** - *sediments* are deposited in layers
- **consolidation** - the conversion of loose sediments into hard *sedimentary rocks*
- **deep burial** - heating and pressure form *metamorphic rocks*
- **melting** - rocks melt to form *magma*
- **cooling** - crystallisation of magma to form *igneous rocks*
- **uplift** - huge forces lift rocks slowly to the surface.

The products of the rock cycle are sediments, sedimentary rocks, metamorphic rocks, magma and igneous rocks.

It is important for the pupils to grasp that these processes involve many different timescales, from hours and days on the one hand to millions of years on the other. It is helpful to divide the spectrum into two categories:

- **Short timescales** measured in minutes/hours/days/years: surface processes such as transport of materials in landslides/flash floods/rivers/glaciers; the eruption and cooling of lava.
- **Long timescales** measured in millions of years: deeper processes such the burial and consolidation of sediment to form sandstone, the formation of a mountain belt, the cooling of magma in large intrusions.

The rock cycle has acted continuously but very slowly since the Earth's formation approximately 4600 million years ago. The rock cycle is driven by plate tectonic forces; see Appendix E.

The activity initially asks pupils to describe each of the rock specimens, with a few prompts to help them on their way. Pupils should be encouraged to interpret what they see in terms of the processes by which the rocks have been formed. The more able pupils may then be able to suggest names for some or all of the specimens. Less able pupils could be encouraged by being given a list of names (± simple definitions) from which they could select appropriate names; simple descriptive tables upon which pupil guidelines could be based are given in Units 1, 5 and 6.

Answers to the questions on the student sheet

1. The teacher will need to match the descriptions against the index number provided for the rock. Grain size is an important characteristic and ought be mentioned in all descriptions.

2. The presence of interlocking crystals usually indicates an igneous rock (see flowchart p.**73**). A reaction with dilute hydrochloric acid indicates that the rock contains a carbonate and is either sedimentary or metamorphic in origin. Granite and basalt solidified from magma. Limestone, sandstone and mudstone settled at the bottom of the sea. Slate was cooked and squeezed deep in the crust.

3. It is not intended that pupils remember long lists of rock names; you may wish to use the flowchart on p.**73** for simple rock identification.

4. The major differences would be grain size and colour. Igneous rocks such as basalt would have been formed as fine grained extrusive rocks from volcanoes on land and on the sea floor. Coarse grained igneous rocks such as granite would have been formed deep underground in the intrusions shown under the mountain range on the left hand side of the diagram.

5. The major differences would be grain size, shape of fragments and colour. Medium grained sedimentary rocks, such as sandstone, would have formed close to the mountain ranges, either on land or in the sea. Fine grained sedimentary rocks, such as chalk (one variety of limestone) and mudstone, would have been formed further from the coastline.

6. Metamorphic rocks, such as slate, would form within the mountain belt on the left hand side of the diagram where plate forces and high temperatures are located.

Answers to the questions on the homework sheet

1. The answer should include the weathering of granite to break it up into smaller fragments; erosion which continues the breaking up process and sorts the larger particles from the smaller ones, deposition and compaction. More able pupils could state that the granite

contains three minerals, quartz, feldspar and mica, and that chemical weathering converts feldspar and mica into clay minerals.

2. Mudstone is the parent rock for slate. When mudstone is subjected to high pressure during a period of mountain building, the high pressure changes mudstone to slate.

3. Rock cycle diagram.

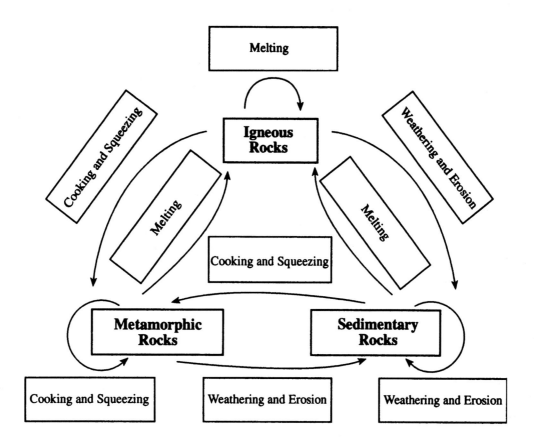

7. Recycling the crust - The Rock Cycle

Student Sheet

The ever-changing Earth

Carefully study the rock specimens provided. The set includes at least one igneous rock, at least one sedimentary rock and at least one metamorphic rock.

1. Describe each rock carefully in your notebook or on the sheet provided:
   — is it *coarse-grained* or *fine-grained*?
   — can you see well-shaped *crystals* in it?
   — do you recognize any *minerals* in the rock (e.g. quartz, mica...)?
   — does it have a 'platy' or 'flattened' appearance?
   — can you see any *fossils* in it?
   — what *colour* is the rock?
   — what other features do you notice?

2. What do these observations tell you about the *formation* of each rock type? Did it:
   — solidify from a magma?
   — settle at the bottom of the sea?
   — get cooked and squeezed deep in the crust?

   Remember to give your evidence and use your conclusions to divide the specimens into sedimentary, metamorphic and igneous groups.

3. Can you suggest appropriate rock-names for any of the specimens? (Use the guidelines provided by your teacher).

7. Recycling the crust - The Rock Cycle

Block diagram of part of the Earth's crust

4. Look at the rocks you think are *igneous*. If you have more than one, in what ways are they different?

   On the block diagram of part of the Earth's crust, identify the places where each of your igneous specimen(s) might have formed. Indicate why you think so.

5. Look at the rocks you have said are *sedimentary*. If you have more than one, how are they different from each other?

   On the block diagram, identify the places where each one might have been formed, giving your reasons.

6. Where might your metamorphic rock(s) have been formed in the block diagram? Explain your reasoning.

Homework Sheet

Recycling the crust

During your lesson you looked at six rocks. One was an igneous rock, called granite. Another was a sedimentary rock called mudstone.

Question 1: **How may a granite provide the raw material for a mudstone?**

One of the other rocks you looked at was a metamorphic rock called slate.

Question 2: **How is slate related to mudstone?**

You have begun to look at an idea that geologists call the Rock Cycle. This shows how the rocks on the surface of the Earth and in the crust can be changed from one form to another. Three processes are important here:

a. weathering and erosion

b. heating and squeezing (application of heat and pressure)

c. melting

Use these processes to label each of the arrows in the Rock Cycle diagram on the following page. Two have been done to get you started; note that you should use each set of words three times.

7. Recycling the crust - The Rock Cycle

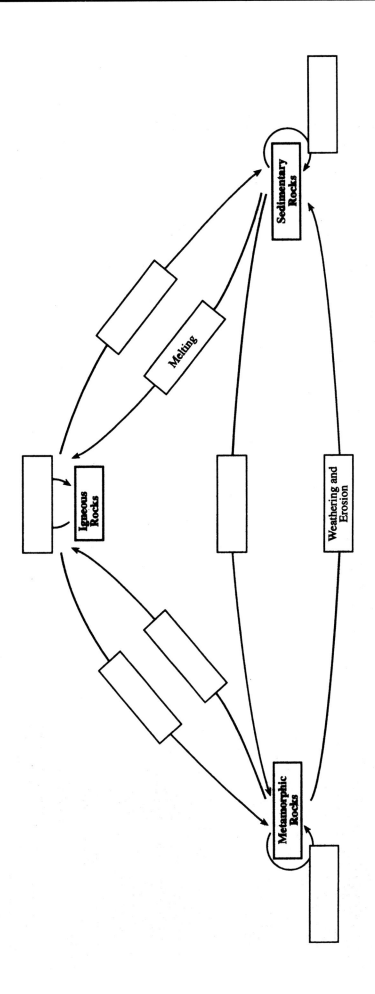

Appendices

A - Concept development

B - Technician's guide

C - Resources

D - Glossary

E - An introduction to plate tectonics

Appendix A - Concept development

Each unit concentrates on one major concept and the activities have been designed to be stimulating and to prompt pupils to ask questions. There will be a range of appropriate learning objectives for each concept depending upon the previous assessment of the group of pupils.

The table below lists the major concept in each unit and gives a range of learning objectives based upon:

what *all* pupils *must* know or understand

what *most* pupils *should* know or understand

what *some* pupils *might* know or understand.

Molten Earth Investigating the cooling/solidification of molten material.	All	Molten material solidifies to form igneous rocks that contain crystals.
	Most	Different cooling rates produce igneous rocks which contain different sizes of crystals.
	Some	Understand that the characteristics of an igneous rock depend upon the depth at which it crystallized and its chemical composition.
'All that glisters . . .' Identifying minerals by their physical and chemical properties.	All	Calcite reacts with hydrochloric acid to produce carbon dioxide gas.
	Most	Be able to distinguish between quartz and calcite.
	Some	Understand that the physical properties of a mineral are determined by its chemical composition and atomic structure.
Whether it weathers Investigating chemical and physical weathering.	All	Weathering breaks rocks down to smaller fragments.
	Most	Changes in temperature and chemical reactions with water cause rocks to break up.
	Some	Understand the processes of chemical and physical weathering.
Shake, rattle and roll Investigating erosion.	All	Rocks are changed to sediment.
	Most	Be able to link the amount of work required to particle size of sediment.
	Some	Be able to relate particle shape and size to the distance/time of transportation.
Sticking together Investigating the formation of sedimentary rocks.	All	Sediment becomes sedimentary rock.
	Most	The strength of a sedimentary rock depends upon the cement.
	Some	Sedimentary rocks are formed by the compaction and cementation of sediments (and may contain animal and plant remains).

Appendix A - Concept development

Feeling the strain An introduction to metamorphic rocks.	All	Understand that mudstone changes to slate when squeezed and heated inside the Earth.
	Most	Understand that some rocks are changed by high temperatures and pressure, e.g. *limestone recrystallises to form marble.*
	Some	The degree of alteration is due to three factors: (1) temperature change (2) amount of pressure (3) the timescale over which (1) and/or (2) operate.
Re-cycling the crust The rock cycle involves sedimentary, metamorphic and igneous processes which take place over different timescales.	All	Can identify a rock as being sedimentary, metamorphic or igneous.
	Most	Be able to describe the processes by which a granite provides the material for a mudstone and the mudstone, in turn, changes into slate.
	Some	Be able to draw and label a diagram showing the three rock groups and the processes which bring about conversion from one rock group to another and indicate the timescales involved.

The criteria for the assessment of pupils' understanding should be the observable outcomes of activities related to the learning objectives. We have made no attempt to match the learning objectives to level descriptions. The 1995 Science Order gives two examples of level descriptors which are related to the Earth science component.

Level 6 They relate changes of state to energy transfers, in contexts such as the formation of igneous rocks.

Level 7 They apply their knowledge of physical and chemical processes to explain the behaviour of materials in a variety of contexts, such as the way in which natural limestone is changed through the action of rainwater, or ways in which rocks are weathered.

For teachers who wish to use an 'end of topic' test the following questions from SCAA Science Tests are still appropriate:

1993 Tier 3-6, Paper 3, questions 1,2,11; Tier 5-8, question 11.

1994 Tier 3-6, Paper 2, questions 4,5.

Appendix B - Technician's guide

1. Molten Earth

- Beaker - 250 cm³
- Boiling tube
- Gauze
- Bunsen burner
- Tripod
- Safety mat
- Safety glasses
- Microscope slides
- Salol (phenyl salicylate)
- Igneous rocks - 2 types are needed and these should show a clear contrast in crystal size. In one crystals should be clearly visible. **Granite** is recommended. The other should have much smaller crystals which are either not visible or barely distinguishable, such **basalt**. The specimen list in the resource pack may help in the acquisition of specimens.
- Access to a refrigerator
- Magnifier or hand lens

2. 'All that glisters . . .'

- Piece of copper foil
- Steel nail
- Dilute hydrochloric acid in dropper bottles
- Mineral specimens (quartz, calcite, galena, pyrite, feldspar and mica are recommended)
- Information Sheet 1: identifying minerals
- Information Sheet 2: properties of some important minerals

The investigation involves the pupil in identifying mineral specimens according to the properties listed in the Student Information Sheets 1 and 2. Each pupil should attempt to identify at least four different minerals. The choice will depend on availability, but the four should include *quartz* and *calcite* (very common minerals that will be useful for subsequent exercises) and at least one mineral with a metallic lustre (e.g. *pyrite*, *galena*).

3. Whether it weathers

Activity A (per group)
- Rock samples: sandstone, slate
- Beakers - 250 cm³
- Tongs
- Crushed ice
- Safety goggles

Activity B
- Measuring cylinder
- 'Mixer' bottle
- Polythene bag

Activity C
- Rock samples: granite, limestone
- Carbonated mineral water
- Glass bottles

4. Shake, rattle and roll

Required by each group of pupils

- 20 pieces of brick. Samples of crushed rock can be used as well or instead of brick, suitable examples are granite, sandstone and slate.
- Large plastic container
- Tray to collect fine residue when sieving
- Colander
- Access to a balance
- Graph paper)
- Student sheet)   per pupil
- Homework sheet)

5. Sticking together

- Piece of sandstone
- Petroleum jelly
- Damp sand
- Powdered clay
- 'Plaster of Paris'
- Syringe

Appendix B - Technician's Guide

The sand should be fine in size, and only slightly damp. You may find it best to make up the mixtures for the pupils. Plastic syringes suitable for the activity are cheaply available from most suppliers. ***You need to cut the end off completely*** with a saw. ***Do not*** be tempted to use discarded syringes from doctors/hospitals, etc. A cheap alternative to a plastic syringe is a piece of plastic tubing and a dowel rod.

6. Feeling the strain
❑ Playdough

Playdough is a cheap, clean alternative to other materials. The required ingredients are as follows:

> Flour - 2 cups
> Water - 2 cups
> Salt - 1 cup
> Oil - 2 tablespoons
> Cream of Tartar - 2 teaspoons
> Colouring

Mix the ingredients together and heat over a low flame. Stir continuously until a ball forms. Allow the mixture to cool and then knead. This makes enough playdough for 4-5 groups of pupils.

7. Recycling the crust

The pupils will need access to 5-6 rock specimens per group, including at least one igneous (i), one sedimentary (s) and one metamorphic (m) specimen (to facilitate assessment, each should be marked with a unique index letter or number).

A suitable list would be:

> granite (i) basalt (i) slate (m)
> limestone (s) sandstone (s) mudstone (s)

❑ Hand lens
❑ Binocular microscope

Appendix C - Resources

1. Sources of information

The following references contain materials/ideas/information which may be of use when implementing the Earth Science component of Key Stage 3.

1. Understanding Geology by D. Webster (ISBN 0 05 003664 5), Oliver & Boyd.

   An excellent GCSE text, which provides a thoroughly lively, practical and comprehensive introduction for teachers.

2. Science of the Earth — 11-14 units, published by ESTA., available from ASE and Geo-Supplies.

   Published as 3-unit packs, each unit occupying a double lesson. Consists of pupil worksheets and guidance for the teacher in setting up experiments.
   Currently available:
   1. Groundwork — Introducing Earth Science
   2. Moulding Earth's surface
   3. Second-hand rocks
   4. Magma
   5. Hidden changes in the Earth
   6. Power Source - oil and energy
   7. Steps towards the rockface - introducing fieldwork
   * Will my gravestone last? (Available as a photocopy from Geo-Supplies only)

3. Clues from the Rocks - Basic Earth Science, Shell Education Service.

   A teacher resource booklet.

4. The Story of the Earth (ISBN 0 11 310023 X)

   Volcanoes (ISBN 0 11 310027 2)

   Rock Solid (ISBN 0 11 310041 8)

   All published by The Natural History Museum. Colourful and lavishly illustrated, but the text can be heavy going for younger pupils. Ideal for the school library.

5. Rocks Around You / Rocks Around You Teachers' Notes. Published by Hobsons Publishing, plc.

   A useful guide to the use of geological materials in society with ideas for practical work.

6. Longman Illustrated Dictionary of Geology (ISBN 0582 55549 3)

   A very useful resource for teachers — all schools should have one.

7. Earth - Eyewitness Science (ISBN 0 7513 1044 1), Dorling Kindersley

   Colourful, will interest most pupils, ideal for the school library.

8. Rocks and Minerals - Eyewitness Handbook (ISBN 0 86318 810 9). Dorling Kindersley

   A good recognition guide for rocks and minerals.

9. Exploring Earth Science CD Rom, Attica Cybernetics.

   A comprehensive collection of colour pictures, audio and video clips; includes ESTA Science of the Earth Units, provides support for the three main subject areas of Geology, Satellites and Space Science.

2. Specimens

(a) Recommended specimens

Igneous rocks

1. Granite
2. Basalt
3. Obsidian

Sedimentary rocks

1. Conglomerate
2. Sandstone (desert)
3. Shale
4. Mudstone
5. Chalk (finer grained variety of limestone)
6. Limestone (shelly)

Metamorphic rocks

1. Slate
2. Marble

Minerals

1. Quartz
2. Calcite
3. Pyrite
4. Galena
5. Mica
6. Feldspar
7. Gypsum

The lists above are by no means comprehensive. However, the presence of these specimens in a school would allow a wide variety of activities to be undertaken in order to implement Key Stage 3.

(b) Acquisition of specimens

Suitable specimens can be readily purchased (see Suppliers, Section 3). Rock specimens are normally sold by size. For class use it is recommended that individual specimens measure 5cm by 5cm or larger. Some suppliers deal in bulk, selling by weight. This is often more economical and you will need bulk specimens for 'Shake, rattle and roll!'

Tiling shops sell polished slabs of a variety of rocks, approximately 30 cm by 30 cm. It is worthwhile purchasing a granite slab to show the crystalline nature of an igneous rock.

Specimens can be acquired more cheaply but it may involve you in some time and effort. Stone masons often have 'off-cuts' which make excellent teaching specimens, if you can identify them. Many museums have a school-loan system, but, of course, the specimens must not be damaged in any way. Sixth Form Colleges, Colleges of Higher Education and Universities may have materials available and can be a valuable source of assistance. It may be possible to collect specimens from field trips, although this will depend on geographic locality.

For those of you who wish to identify your own specimens, an appropriate flowchart is provided (opposite).

3. Suppliers of equipment and materials

1. Northern Geological Supplies, 66 Gas Street, BOLTON BL1 4TG

   Wide range of specimens and equipment available.

2. OFFA Rocks, Lower Hengoed, OSWESTRY, Shropshire SY10 7AP

   Wide range of mineral, rock and fossil specimens. Bulk samples available.

3. Richard Taylor Minerals, Byways, 20 Burstead Close, COBHAM, Surrey KT11 2NL

   A wide range of rocks and minerals.

4. Gregory, Botley and Co., 8-12 Rickett Street, LONDON SW16 1RU.

   Collection of rocks, minerals and fossils.

5. GEOU, Department of Earth Sciences, The Open University, Walton Hall, MILTON KEYNES MK7 6AA.

   A wide range of fossil casts and rocks.

6. Landform Slides, Mr. K. Gardner, 38 Borrow Road, Oulton Broad, LOWESTOFT NR32 3PN.

   Wide range of colour transparencies.

7. MJP, Freepost, P.O. Box 23, ST. JUST TR19 7JS

   A range of apparatus, computer software, slides and OHP transparencies.

8. Geo-Supplies Ltd., 16 Station Road, Chapeltown, SHEFFIELD S30 4XH

   A large range of books, equipment and specimens.

9. Philip Harris, Lynn Lane, Shenstone, LICHFIELD WS10 0EE

   Instruments, equipment and specimens

10. Griffin and George Ltd., Bishop Meadow Road, LOUGHBOROUGH LE11 0RG

    Instruments, equipment and specimens of rocks, minerals and fossils.

4. Useful addresses

1. **British Gas Film & Video Library,** Park Hall Road Trading Estate, London SE21 8L

   Some useful videos for loan or purchase.

2. **BP Educational Service,** P.O. Box 30, Blacknest Road, Alton, Hampshire GU34 4PX

   A wide range of books, booklets, computer and slide packs, videos and films. Some materials free, others for loan or purchase.

Appendix C - Resources

FLOWCHART FOR SIMPLE ROCK IDENTIFICATION

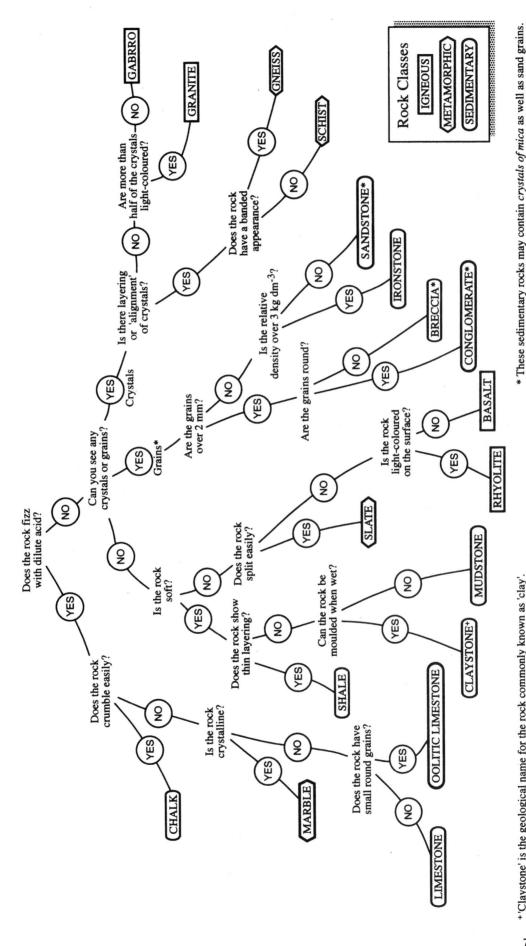

* These sedimentary rocks may contain *crystals of mica* as well as sand grains.

+ 'Claystone' is the geological name for the rock commonly known as 'clay'.

Appendix C - Resources

3. **Earth Science Teachers' Association,** c/o The Geological Society, Burlington House, Piccadilly, LONDON W1V 0JU

   The Association publishes a journal 'Teaching Earth Sciences' quarterly, and is involved in curriculum development.

4. **ICI Video Tapes,** Argus Film & Video Library, 15 Beaconsfield Road, LONDON NW10 2LE

   A limited selection of videos for purchase.

5. **Royal Holloway, University of London,** Department of Geology, EGHAM, Surrey TW20 0EX

   The College runs two projects of interest to schools:-

   (i) **Earth Sciences Resources Centre** - has some resources available for loan and can offer advice/assistance.

   (ii) **RH Earth Science** - is a curriculum development and INSET project.

6. **Shell Education Service,** Shell UK Ltd., Shell-Mex House, Strand, LONDON WC2R 0DX

   Resource material, wallcharts, booklets and other aids, many available free.

7. **Esso Information Service,** PO Box 695, SUDBURY, Suffolk CO10 6YM

   Resource material and wallcharts available free. Videos available on loan.

8. **The RTZ Corporation plc,** (Educational resources service), 6 St. James's Square, LONDON SW1Y 4LD

   Resource material, mineral samples and booklets available free. Videos available on free loan.

9. **British Geological Survey,** Kingsley Dunham Centre, Keyworth, NOTTINGHAM NG12 5GG

   Books, maps, wallcharts.

10. **The Geological Society,** Burlington House, Piccadilly, LONDON W1V 0JU

    Booklets and wallcharts, advice available from the Education Officer.

Appendix D - Glossary

Abrasion
the wearing away of a surface when particles are dragged over or hurled against it.

Attrition
the mutual wearing away of the surfaces of *rock* particles as they collide in the course of being moved by water or wind.

Aureole
a zone of *metamorphic rocks* surrounding an *igneous intrusion*, whose metamorphism has been caused by heat derived from the intrusion (see *contact metamorphism*).

Basalt
a dark *igneous rock* made up of fine (less than 1 mm in size) crystals. Basalt consists of the minerals pyroxene and feldspar. The dark colour is due to the large content of pyroxene.

Batholith
a large scale *igneous* intrusion, tens or hundreds of kilometres across; usually the product of multiple intrusions of granite magma.

Bed (bedding)
distinct layer (layering) in sedimentary rocks, representing successive episodes of deposition. See also *strata*.

Conglomerate
a *sedimentary rock* made up of large rounded fragments (exceeding 2 mm) cemented together by a finer matrix.

Contact metamorphism
metamorphism arising due to heat transferred to *rocks* by an *igneous intrusion*. See *aureole*.

Crust
the outermost, chemically distinct layer of the solid Earth. Divided into *continental crust* (rich in silica) and *oceanic crust* (rich in iron/magnesium).

Dolerite
a dark *igneous rock* consisting of medium sized crystals (1 - 5 mm), made up of the same minerals as *basalt*.

Dyke
a near vertical sheet-like *intrusion* of *igneous rock* (commonly 0.5—5.0 m thick) cutting across other *rock* units.

Erosion
the physical removal ('transport') of *rock* fragments; agents of erosion include water, wind and ice. See *attrition* and *abrasion*.

Extrusive
refers to *igneous rocks* that have erupted on to the Earth's surface.

Gabbro
a dark *igneous rock* consisting of coarse crystals (larger than 5 mm); made up of the same minerals as *basalt* and *dolerite*.

Glass
a volcanic *rock* formed by *magma* of any composition that has been cooled too quickly to form crystals. Shows curved, shiny fracture surfaces like broken bottle-glass, but is usually black and opaque.

Granite
a light-coloured *igneous rock* consisting mainly of large (larger than 5 mm) crystals of quartz and feldspar. Other minerals present may include mica (light or dark), hornblende or tourmaline. Very viscous type of magma.

Hardness
the ability of a *mineral* to resist scratching. Usually expressed on a numerical scale from 1 (talc) to 10 (diamond): Mohs' scale.

Igneous
refers to crystalline or glassy *rocks* that have solidified from the molten state (*magma*).

Intrusion
a body of *igneous rock* formed when *magma* is emplaced into the crust. Types of intrusion include *batholiths*, *dykes* and *sills*.

Intrusive
refers to *igneous rocks* that have crystallized within the crust.

Lava
magma that erupts on the surface as a flowing liquid; a solid *rock* formed by this process.

Limestone
a *sedimentary rock* made up predominantly of the carbonate minerals, usually calcite ($CaCO_3$). Most limestones are made up of the remains of living creatures (e.g. chalk), but some consist of inorganically precipitated carbonate.

Appendix D - Glossary

Lithification
refers to various processes by which loose *sediment* is converted into hard *sedimentary rock*.

Lithosphere
the rigid outer shell of the Earth forming the tectonic plates. Consists of the crust and uppermost part of the underlying mantle.

Magma
rock material in a molten or partially molten state.

Mantle
that part of the Earth's interior situated between the core and the crust.

Marble
a *metamorphic rock* made up of crystals of calcite or other carbonate minerals; a metamorphosed limestone. *Caution: stonemasons may apply the term to any decorative rock, regardless of how it formed.*

Metamorphic
rocks formed by alteration of pre-existing rocks by the effects of heat and/or pressure. Generally metamorphism involves solid-state recrystallization leading to the formation of new minerals.

Mineral
a natural substance having a well-defined, uniform chemical composition and crystal structure.

Mudstone
a *sedimentary rock* with fine particles (less than 0.06 mm) which is structureless, unlike *shale* which shows lamination.

Obsidian
a glassy *volcanic igneous rock*; its composition is usually similar to *granite*.

Plate
a rigid shell forming part of the Earth's lithosphere.

Plate tectonics
The theory describing the dynamics of the rigid outer layer of the Earth.

Regional metamorphism
metamorphism due to widespread heating, compression and deformation in a developing mountain belt such as the Alps or Himalayas.

Rock
to the geologist any mass of mineral matter which forms part of the Earth's crust is a rock, incuding loose sediment as well as hard rock. The popular meaning of the word regards rock as something hard that has to be removed by blasting; this is the meaning we have used in this book.

Sandstone
a *sedimentary rock* made up of sand fragments (usually 0.06—2.00 mm in size).

Sediment
natural debris, such as *rock*, mineral or shell fragments, deposited after transportation by wind or water but not lithified into a hard rock.

Sedimentary
rocks that are formed by the *lithification* of *sediment*.

Shale
a *sedimentary rock* with fine particles (less than 0.06 mm) showing laminations, which is fissile (tends to split into thin sheets parallel to bedding).

Sill
a sub-horizontal sheet-like *igneous intrusion*, commonly intruded between, and parallel to, *sedimentary* strata.

Slate
a very fine-grained *metamorphic rock* that splits very easily.

Strata
collective term for a series of *beds*.

Subduction
the passage of a tectonic *plate* down into the deeper mantle.

Subduction zone
an inclined surface down which a tectonic plate is subducted. Marked by earthquake foci and often a band of volcanoes directly above.

Volcanic
refers to the products of an erupting volcano.

Weathering
the processes by which rocks are decomposed *in situ*.

An introduction to plate tectonics

Supplementary information for the teacher

In 1915, Alfred Wegener wrote a book entitled 'The Origin of the Continents and Oceans'. In it, he proposed the theory of *continental drift*, citing several lines of evidence including the fit of conjugate continental margins, records of past climates, crustal structure and the distribution of plants and animals both past and present. As Wegener himself wrote: *'It is just as if we were to refit the torn pieces of a newspaper by matching their edges, and then check whether the lines of print run smoothly across. If they do, there is nothing left but to conclude that the pieces were in fact joined this way'*.

Contemporary reaction to Wegener's ideas was hostile. Geophysicists considered the Earth's crust too rigid for lateral movement. Views concerning Earth history were dominated at the time by the notion that the Earth was becoming progressively cooler, and crustal deformation was simply the result of thermal contraction.

Wegener's views have been completely vindicated in the last few decades. Very precise satellite-ranging experiments have demonstrated that the continents really are moving relative to each other (New Scientist, 31st May, 1984, p.6). Current opinion explains this movement in terms of *plate tectonics*. It is clear from magnetic and seismic studies that the outermost layer of the solid Earth consists of a mosaic of about a dozen large, solid plates between 50 and 200 km thick, which constitute what is called the *lithosphere**. The plates float on a more ductile interior layer (the *asthenosphere*).

From the study of ocean-floor magnetic anomalies, it is clear that plates move across the Earth's surface at velocities of a few centimetres per year. Plates move outwards from *mid-ocean ridges* (such as the Mid-Atlantic Ridge). A mid-ocean ridge is therefore a zone of accretion at which new lithosphere is constantly being added to the trailing edge of the plate as a result of igneous activity (what is called a 'constructive plate margin'). The consequent increase in surface area is compensated at the *ocean trenches*, where lithospheric plates are diverted back into the interior of the Earth ('destructive plate margins'). Seismologists can track the downward progress of the plate as an inclined zone of earthquake foci, extending from the trench deep into the mantle, usually at an angle of 30-60°. Such *subduction zones*, as they are called, are also the source of intense and often violent volcanic activity, which builds an arcuate chain of volcanic islands (*island arcs*) directly above the subduction zone. The tendency for

---

* *Confusion often arises in school science books between the terms 'crust' and 'lithosphere'. The crust (varying from 7 to 50 km thick) is distinguished from the mantle beneath it on the basis of chemical composition (and the consequent difference in seismic velocity). The plates forming the coherent outer layer of the Earth, on the other hand, are distinguished by their relative rigidity compared to the more ductile layer below. This rigid shell comprises not just the crust but also the uppermost part of the mantle beneath. When the parameters of plate tectonics first became clear, geologists coined the term 'lithosphere' to identify this rigid outer spherical shell, of which the individual tectonic plates form a part. Unfortunately the same word is used by geographers in a different sense, to represent* all *the solid, rocky parts of the Earth (as distinct from the atmosphere, hydrosphere, etc.)*

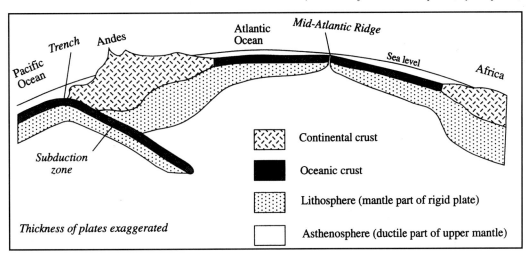

Thickness of plates exaggerated

- Continental crust
- Oceanic crust
- Lithosphere (mantle part of rigid plate)
- Asthenosphere (ductile part of upper mantle)

plates to sink along subduction zones seems to be a major driving force of plate movement.

Some plates are covered entirely by oceanic crust (e.g. the Pacific Plate), whereas others include both oceanic *and* continental crust (such as the African Plate and the Eurasian Plate). Plates are found to be thicker beneath the continents than under the oceans. The continental crust has a lower density than oceanic crust: the elevation of the continents above sea level is due to their buoyancy.

Tectonic plates act like conveyor belts: a continent will be carried along passively in the direction in which the plate of which it forms a part happens to be moving. When a continent arrives at a subduction zone, however, subduction tends to cease as the continent has too low a density to be carried down into the Earth's interior. Some mountain belts like the Alps and Himalayas are zones where two continents, carried toward each other by convergent plate motions, have collided above a subduction zone. The collision accounts for the enormous lateral stresses during mountain-building (*'orogenesis'*), to which intensely folded and deformed strata of the Alps testify. A mountain belt may also develop above a subduction zone where an oceanic plate is subducted beneath a continental margin, as has happened under the Andes. In either circumstance, it is common for deep granite intrusions to be formed.

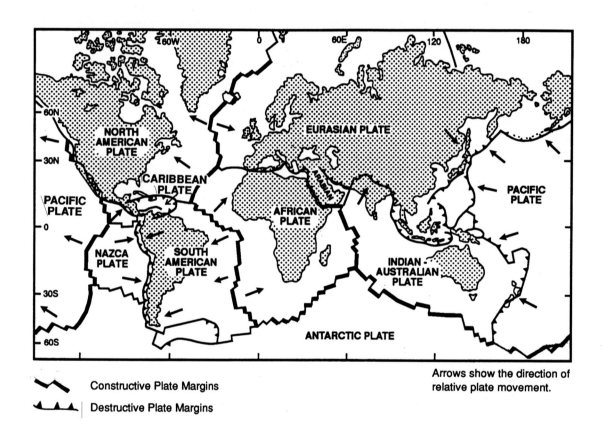

NOTES

NOTES

NOTES

NOTES